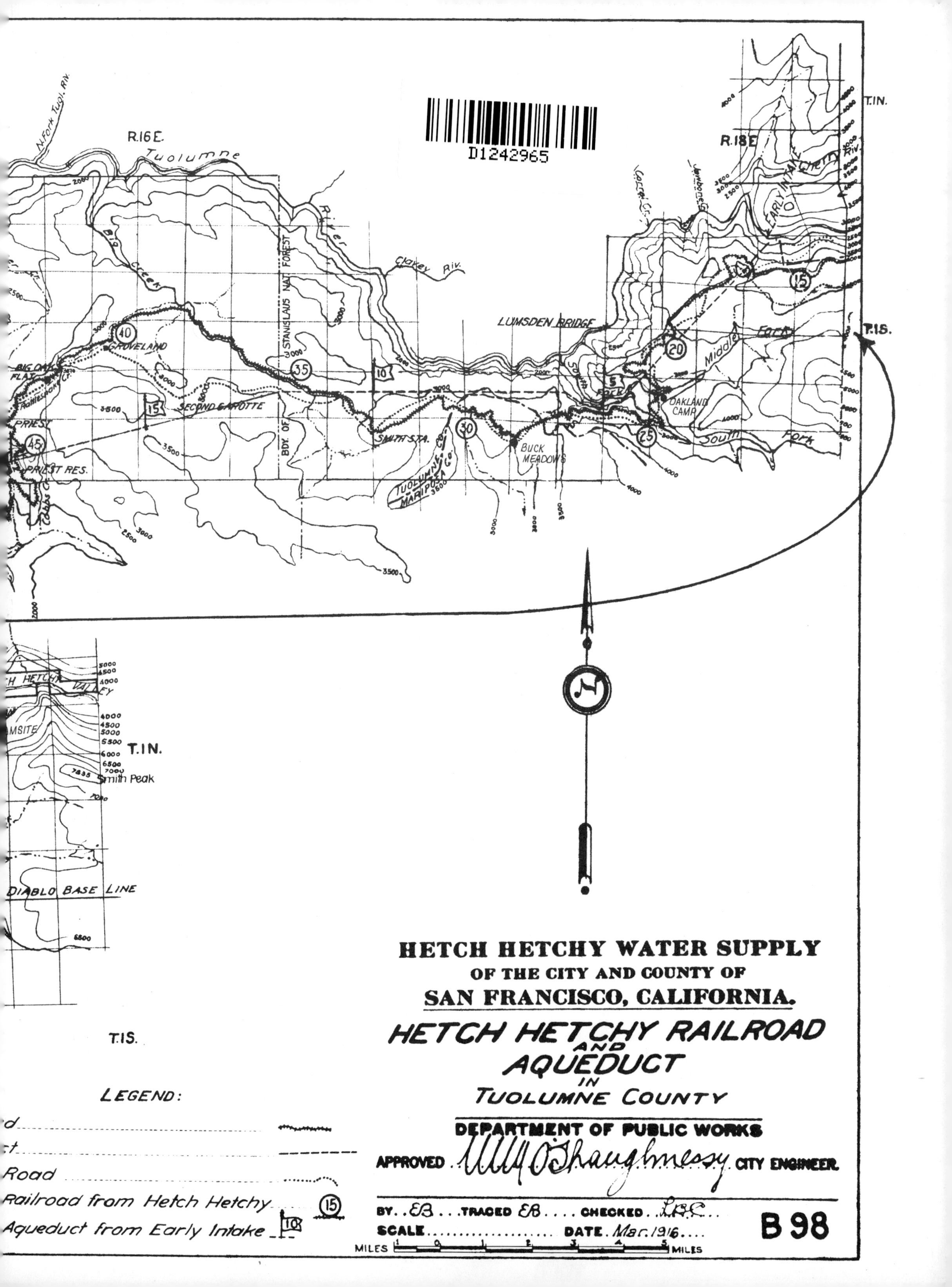
D1242965
R.16E.
Tuolumne
River
N. Fork Tuol. Riv.
Big Creek
Clavey Riv.
STANISLAUS NAT. FOREST
BDY. OF
R.18E.
Cherry Riv.
EARLY INTAKE
LUMSDEN BRIDGE
Middle Fork
South Fork
OAKLAND CAMP
GROVELAND
BIG OAK FLAT
Rattlesnake Cr.
SECOND GAROTTE
PRIEST
PRIEST RES.
SMITH STA.
BUCK MEADOWS
TUOLUMNE CO.
MARIPOSA CO.
T.1N.
T.1S.
HETCH HETCHY VALLEY
Smith Peak
DIABLO BASE LINE
HETCH HETCHY WATER SUPPLY
OF THE CITY AND COUNTY OF
SAN FRANCISCO, CALIFORNIA.
HETCH HETCHY RAILROAD
AND
AQUEDUCT
IN
TUOLUMNE COUNTY
DEPARTMENT OF PUBLIC WORKS
APPROVED M.M. O'Shaughnessy CITY ENGINEER.
BY. EB TRACED EB CHECKED L.B.C.
SCALE DATE Mar. 1916
B 98
MILES 1 0 1 2 3 4 5 MILES
LEGEND:
Road
Railroad from Hetch Hetchy 15
Aqueduct from Early Intake 10

Hetch Hetchy and its Dam Railroad

20
5
PAUL TWINE

Hetch Hetchy and its Dam Railroad

The story of the uniquely equipped railroad that serviced the camps, dams, tunnels and penstocks of the 20-year construction project to bring water from the Sierra to San Francisco.

by Ted Wurm

BERKELEY  CALIFORNIA

HETCH HETCHY AND ITS DAM RAILROAD

Printed and bound in the United States of America

Library of Congress Catalog Card No. 73-87231

ISBN 0-8310-7102-8

Published by Howell-North Books
1050 Parker Street, Berkeley, California 94710

CONTENTS

Preface and Acknowledgments 6
Introduction 9
1 San Francisco's Early Water Supply 13
2 The Twelve-Year Fight 17
3 Tuolumne County Background 29
4 Starting the Big Job 41
5 Building Hetch Hetchy Railroad 47
6 We Have a Railroad! 63
7 Preliminary Construction 73
8 The Big Dam at Hetch Hetchy 85
9 A Mountain Tunnel Gets Started 105
10 Mountain Division Tunnel Completed 119
11 Life in Boomtown Groveland 127
12 Groveland Swings! 135
13 Rail Lines All Over the Place 139
14 Gasoline Alley 151
15 Steam Train Operation 159
16 Excursions, Wrecks and Worms 169
17 Priest, Moccasin and Disaster 179
18 Westward by Tunnel and Pipeline 193
19 Longest Tunnel in the World 209
20 Plentiful Water at Last 229
21 Hetch Hetchy Railroad Lives On 241
22 Sierra Railway Helps to Raise the Dam 251
23 End of the Line Heralds a New Era 263
Appendix 1 Lives Lost in Hetch Hetchy Construction . . 284
Appendix 2 Roster of Railroad Equipment 286
Bibliography 287
Index 290

PREFACE AND ACKNOWLEDGMENTS

In the beginning this was to be a simple story about the building, operation, and fading out of a spectacular mountain railroad. One day in Groveland, California, in August 1964, I made inquiries about anyone who might have worked on the Hetch Hetchy Railroad. It was my good fortune to be directed to Ernie Beck, who supplied stories and pictures of the railroad and gave me the names of several other former railroaders living in the area. One person led to others, and by the end of that year I had talked to 37 people who had worked on or had personal knowledge of Hetch Hetchy.

However, I found myself unable to keep them to the subject of the railroad. A larger theme imposed itself and gradually took over, so that I found emerging a picture of the whole Hetch Hetchy Project. People who had worked just on the railroad operation were relatively scarce. There were many others, though, who had been involved in other phases of the project itself. They shared anecdotes, showed pictures, and their enthusiasm got me completely involved in putting together this story of the entire project.

It is a fascinating story and I found the people who had been involved even more fascinating. My research in 1965 included about 120 interviews in all parts of Central California. In 1966 there were even more, but the years following brought a gradual tapering off, for it was with difficulty that any "new" Hetchy old-timers could be found. Eventually, just over 200 people were consulted, some of them several times, and I was as hooked on the "Hetch Hetchy Spirit" as any of them. This spirit would seem to have developed out of a combination of things: San Francisco's "know-how" attitude, the brazen opposition, the great challenge itself, the atmosphere of the Roaring Twenties and Prohibition, and the exceptionally fine people who went to do the job. So what emerges is the tale of the whole Hetch Hetchy Project – the railroad, of course, the several dams and record-breaking tunnels, a collection of the most interesting machinery and mechanical devices and operations, power and water development, the miners, engineers, trainmen, gamblers, bootleggers, "girls," the "Poison Oakers" of southern Tuolumne County and, over it all, the figure of City Engineer Michael M. O'Shaughnessy, admired "Chief" of Hetch Hetchy.

The City and County of San Francisco, to our good fortune, has not been one to discard willy-nilly any record just because it is old. Rather, the files on Hetch Hetchy are treasures of information and pictures. People in City employ today, especially those in Hetch Hetchy, delight in talking about the project with interested outsiders. I am indebted to many of them for help with this book. These include, in the Hetch Hetchy Department, Oral L. Moore, general manager, Bernard Grethel, Rino Bei, Helene Shields, Elnore Maunder, Jim Eastman, Bill Stahl and Bob Lee. Hetch Hetchy librarians Ardean Heyer, Gertrude Keilin and Roy Pampanin assisted tremendously, as did these men in the engineering section: Wesley Getts, Ernie Orognen, Charlie Reed, Leo Smith and Walter Watson.

Much help was given by people in various offices of the Public Utilities Commission: Ethel Cullen, Jim Leonard, Marshall Moxom, Bill Simons; and the City Engineer's Office under P.U.C.: Tom Beggs, Leo Glick, John Jelincich and Walter

McKenzie. San Francisco Water Department records were made available and the San Francisco Public Library provided help, especially Gladys Hansen in Rare Books and Special Collections. Assistance came from Bill Bendel, retired Municipal Railway official.

Material was consulted in several other libraries: the Berkeley Public Library, the California Historical Society and California P.U.C. libraries, California State Library (Allan Ottley), the Mechanics' Institute Library, Oakland Public Library (Frances Buxton and Erma Davis in the California Room, Miss Rhodes in Science and Industry), and the Pacific Gas & Electric Company, Sierra Club and Sutro libraries. At the University of California in Berkeley, the General Library's newspaper and periodical files proved invaluable; John Barr Tompkins at The Bancroft Library provided illustrations; Willa Baum at the Regional Oral History Office offered advice and assistance; and Jerry Geifer and Mary Stephens at the Water Resources Center Archives dug up many unusual documents and papers. In Sonora the late Ruth Ann Newport provided material from the Tuolumne County Museum and gave constant encouragement.

Many of the maps used in this volume were drawn by George Clare, whose fine work is gratefully acknowledged.

Finally, here are the names of those people who made this book possible, who brought it to life with their personal stories shared with the author. Quite a number have died since being interviewed over the past eight years; I regret that they did not live to see the book they were all so interested in. I hesitate to say that no one has been left out; in such a number, it would be difficult not to miss a few. If this has occurred, I pray the indulgence of anyone not named.

Acknowledgment for help is made to the following people who actually worked on Hetch Hetchy during the construction years, or who are the widows or descendants of those who so labored:

Jim Alexander
Eddie Anderson
Edwin "Red" Anderson
Alfred Andresen
Joe Anker
Charlie Baird
Jim Baker
Max J. Bartell
George F. Bartlett
Al Beck
Ernie Beck
Waldo Bernhard
Jack Best
Jack Bhend
Tommy Breslin
Roy Brooks
Fred D. Brown
Charles "Doc" Burnett
Earl "Cappy" Caplinger
Mike Carmichael
Ray Carne
Joe Cavagnero
Harry F. Chase
Edith Cheminant
Robert Cheminant
Harry Cobden
Burgess Cogill
Tom Connolly
George Connor
Mary Conty
Art Crowley
Maude Crowley
Frank Cummings
Earl Cutting
Ernie De Ferrari
Frank "Butch" De Ferrari
Dr. John P. Degnan
John L. Donaldson
Grace Eckart
Owen Ellis
Charles M. Fanning
Hank Femons
Curtis "Red" Fent
Bill Firmstone
David Fisher
Jack Garrison
Del Gilliam
Pete Golub
Jim Graham
E. B. "Hans" Hansen
Cliff Hanson
Bill Haynes
Grace Helbush
W. W. "Bill" Helbush
Laura Hennessey
Dorothy Hinkson
Earl P. Hope
Phil Hope
Willard James
Leonard "Slim" Jameson
Calvin Jones
Al Kyte
George Laveroni
Gilbert Leach
Emilie McAfee
Louis A. McAtee
Dorel McDowell
Phyllis Cheminant MacLean
Irene Magee
Walt "Eddie" Magee
Mary Male
Eugene "Bud" Meyer
Mary Meyer
Bert Minard
Doug Mirk
Muriel Mirk
Walt Mitchell
Edward P. Morphy
Anna Muheim
John F. Mullen
Herman A. Nagel
James W. Neel
Marie Floyd Neel
Axel "Ole" Olson
Miss Bess O'Shaughnessy
Miss Helen O'Shaughnessy
Miss Mary O'Shaughnessy
Rueben H. Owens
Earl Palmer
Vern Peugh
Eddie Phelan
John "Les" Phelan
Jerry Pickle
Earl Pool
David S. Raggio
Carl R. Rankin
Ernie Rawles
Bill "Busher" Reed
Merle Rodgers
John "Buddy" Ryan
Carl Seward
Al Simpson
Dudley Snider
Roy "Blackie" Stevenson
Leslie W. Stocker
Ottilie Taylor
Mrs. Chas. Coyle Thomas
Donald O. Townsend
Kate Townsend
Theresa Townsend
James H. Turner
Kathy Meyer Venable
Julius Verkuyl
Louise Harwood Waldron
Bill "Red" Wanderer
Lloyd Whaley
Herb Whipp
E. J. "Jack" White
Ray White

Acknowledgment for help with story and pictures is made to these persons, residents or one-time residents of Tuolumne County:

Margaret Corcoran Anker
Louise Goldsworthy Baird
John Bartlett
Lila Phelan Best
Olive Pool Caplinger
Frank Cassaretto
Fred "Rico" Cassaretto
Mrs. Fred Cassaretto
Carlo M. De Ferrari
Mary De Ferrari
Pat Egan
Barney Emerson
Orion and Anne Fenton
Bill and Georgia Gookin
Bernice Rovero Guthals
Catherine Cobden Haight
Mona Hanson
Kay Harms
Robert Hooe
Marvin Hope
Flossie James
Gladys Cobden Jameson
Jacunda Laveroni
Fred Leighton
Gladys Hope Little
Selina Lumsden
Grace Stanley McCaslin
Wilfred and Wilma McCoy
Harvey McGee
Millard C. Merrill
Ken Mirk
F. Mervyn Munn
Velma Nelson
Agnes Palmer
Margaret Conty Pool
Iola Rodgers
Dan and Angie Rovero
Pat Ryan
Jim Segale
Mabel Seward
Emily Magee Sikola
Marguerite Golub Snider
Fred R. Stanley
Harry Stanley
John Turner
James White
Mary De Ferrari White
Susie White
Bernice Laveroni Workman
Les Young

Special thanks are due to the following, neither workers on Hetch Hetchy nor residents of the area involved, who assisted with advice, technical information, photographs, and engineering and historical data:

Dolores Anderson
Al Barker
Lee E. Brooks
Walter C. Casler
Charles O. Clerk
Harre Demoro
Bob and Lorraine Dillon
Stephen E. Drew
Marty Drury
Sherman Duckel
J. T. Easler
Francis Farquhar
Al Fickewirth
Frank Foehr
Bernard Gallagher
G. M. "Curly" Gardiner
Cyril Gilfether
Alvin Graves
Roy D. Graves
Francis Guido
Adolf Gutohrlein
Gerald Harrington
Gladys Mary Harris
Lois Hart
George Henderson
W. C. "Dutch" Hendrick
Douglas Hubbard
John Hussey
Hank Johnston
Holway Jones
Mr. & Mrs. Art Kelleher
Ralph Kerchum
William C. Kluver
Addison Laflin, Jr.
James Law
Frank Male
Vince Martini
Marvin Maynard
Thomas F. Means
Roy V. Meikle
Warren E. Miller
Rick Mugele
David F. Myrick
Bill Pennington
Edward T. Planer
Clifford Plummer
Mr. & Mrs. R. O. Richardson
Douglas S. Richter
E. M. Rogers
F. S. Rolandi, Jr.
Al Rose
Genevieve Rowe
Vernon Sappers
Shirley Sargent
Bob Schlechter
Margaret Schlichtmann
D. J. Shelburne
Charles Smallwood
Stanley Snook
Edna Ellis Speed
Louis L. Stein, Jr.
Joe Strapac
William P. Tuggle
George Turner
Utah Construction Co.
Leonard "Knave" Verbarg
Bert Ward
W. E. Waste
Jeffery Wetmore
White Motor Company
Hugh Wiley
John Woods
James F. Wurm
John D. Wurm
Walter S. Young

It should be pointed out that since the time of the Gold Rush in Northern and Central California, the term "The City" has always meant San Francisco. Similarly, "The Peninsula" refers to that geographical area south of the City, bordered on the east by San Francisco Bay and on the west by the Pacific Ocean. The terms are considered proper names for those areas named, and it is in this manner that the terms will be used throughout this book.

This work is sincerely dedicated to "The Chief," Michael Maurice O'Shaughnessy, and to all those who worked with him and for him in building Hetch Hetchy, and to those who have followed in their footsteps.

Ted Wurm

Oakland
March 1973

INTRODUCTION

The City and County of San Francisco owns and operates a water and power system in which its residents take great pride. The conception and achievement of this system made up an engineering feat of great magnitude, a contribution not only to the welfare of the people but also to the development of San Francisco and the general metropolitan area. Modern urbanization demands ever-increasing supplies of clean, inexpensive water. There are few cities in the world that have a water supply so dependable, of such excellent quality, and adequate to all foreseeable needs well into the next century.

But this was not always the case. Until 1930 the City's water was furnished by a private utility, a company that occasionally ran out of water, that was often unable to supply outlying districts and that charged relatively high rates for the service. The inadequacy of this supply had become evident to the citizens late in the 19th century, and it was brought strongly to their attention at the time of the earthquake and fire of 1906, when the City was largely destroyed for lack of water to quell the flames. A new charter of 1900 had empowered the people of San Francisco to take over or develop their own public utilities. With this mandate, the people looked to the distant Sierra Nevada for water, and they set out to develop and build their Hetch Hetchy Project. This system taps the water of Tuolumne River, flowing out of the snow-capped mountains about 150 miles east of the San Francisco Bay region.

There were problems of rights of way and water rights. There was the problem of financing such a vast project by a municipality as small as San Francisco was at that time. Federal and state funds were not available in those days. There was tremendous opposition from the private water company, which sought to keep its monopoly control of City water supplies, and irrigation districts using Tuolumne water lined up against San Francisco. Also in the fight were the Sierra Club and other conservationists wishing to preserve the mountain areas from "desecration."

The major problem of just getting men and materials into the wild mountain areas had to be overcome by the building of a marvelous railroad, the story of whose quaint operation and unusual equipment fills a major part of this book. The planning and operation of scores of construction camps were challenges that faced City officials who dared tackle such a mammoth project. And these were camps that throbbed with activity during the days of Prohibition – the time when a bootlegger of homemade "jackass," a whorehouse, a gambling joint, or an illegal saloon could be found behind every tree! And most of this construction was in a rugged, heavily forested mountain area, a country quite wild and far removed from the city life most workers were used to.

The genius of City Engineer Michael Maurice O'Shaughnessy, who served also as Chief Engineer of the Hetch Hetchy Project, was largely instrumental in solving the financial problem. By completing revenue-producing facilities first and by carefully extending initial construction over a period of twenty years, O'Shaughnessy was able to keep the work barely within financing capabilities, despite the inflationary period of World War I and difficulties caused by the Great Depression starting in 1929. After an expenditure of $100,000,000, the first water flowed into San Francisco in 1934.

This book attempts to tell the story of that construction job, one of the biggest ever attempted up to that time. It will tell of subsequent additions and improvements – great designs that have continued to enlarge the project, and still essentially based on the original planning. Indeed, the work of subjugating the Tuolumne River and conserving every drop of its water to a useful purpose is still in progress. As expenditures to date are passing the $500,000,000 mark, Hetch Hetchy water is supplied for residential, commercial and industrial use in a 500-square-mile service area comprising San Francisco as well as neighboring communities in most of San Mateo County and parts of Santa Clara and Alameda counties.

The work of thousands of truly dedicated men and scores of hard-working women, beginning more than half a century ago, was responsible for bringing into being this tremendous asset. It is the purpose of this book to show how they did it with the limited but fascinating machinery and tools available at that time, and that remarkable "Spirit of Hetch Hetchy," which perhaps comes alive in these pages.

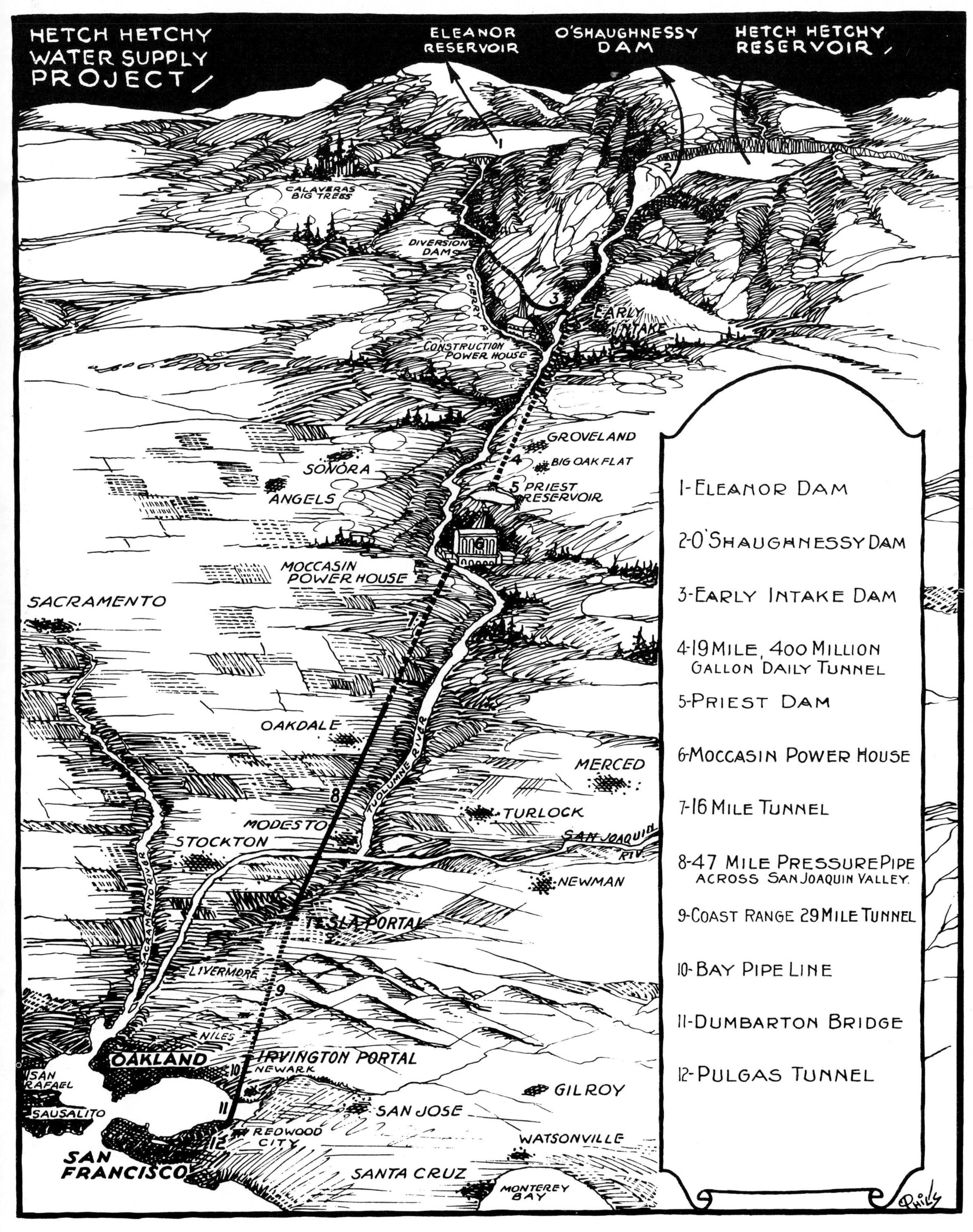

This drawing was made upon completion of initial Hetch Hetchy Project construction in 1934. The water flows all the way to San Francisco by gravity, no pumping being necessary at any point.

San Francisco Water Works began bringing water to the City from distant Lobos Creek in 1858. These two views from the 1860s show part of the five-mile wooden flume as it rounded Black Point (present Fort Mason). From here it was pumped to reservoirs at Francisco Street and at Lombard; outline of the latter is shown at top of the distant hill. The cove is the location of today's Aquatic Park at the foot of Van Ness Avenue. *(–S.F. Public Library)*

The lower view shows a corner of Pilarcitos Reservoir in San Mateo County, with the wooden flume carrying water on its journey into San Francisco. *(–S.F. Water Dept.)*

1

SAN FRANCISCO'S EARLY WATER SUPPLY

Lying at the center of California's coastline, the San Francisco Bay region is classified as semi-arid. The City itself sits at the end of a forty-mile-long peninsula, with salt water on three sides. Therefore, the local fresh-water supply is quite limited. The struggle for it began in 1776, when Franciscan missionaries selected a small stream as the site of their Mission San Francisco de Asís (Mission Dolores today). The Gold Rush of '49 rapidly transformed the sleepy little pueblo of Yerba Buena into the booming City of San Francisco.

A continuing problem from the beginning was to find an adequate supply of water. Some residents were able to dig shallow wells or get water from springs, while others bought theirs from peddlers, who ferried it in barges from Marin County, across the Golden Gate to the north. It is reported that during one period of scarcity people on some of the "water routes" were actually charged one dollar in gold per bucket – the start of another gold rush! Writer Anita Day Hubbard said, "The direct ancestor of the Hetch Hetchy system was a donkey. . . . A humble burro that carried two barrels from which water was doled out to buyers. . . ."

The first water company established in San Francisco was the Mountain Lake Water Co. of 1851, formed to bring water from Lobos Creek and Mountain Lake in the present-day Richmond District of western San Francisco. In 1857 the San Francisco Water Works was organized by John Bensley and Anthony Chabot, tapping Lobos Creek and carrying two million gallons per day to the center of town. On September 16, 1858, water started flowing through five miles of wooden flume hung along the south shore of the Golden Gate, passing through a tunnel at Fort Point. A pumping plant at Black Point, near today's Aquatic Park, raised this water to two storage places on Russian Hill: Lombard and Francisco reservoirs, which are still in use. Small amounts of water were developed elsewhere locally, but these sources were all limited. Operators of the Spring Valley Water Works, organized in 1860 around a small spring near Washington and Mason streets, realized this and pressed south into neighboring San Mateo County, where in 1862 they began development of what is now the Peninsula System of the San Francisco Water Department. Spring Valley absorbed the San Francisco Water Works in 1865.

A reservoir constructed by damming Pilarcitos Creek, about ten miles south of the City line was the first completed unit of Spring Valley's Peninsula works. From here a daily supply of two million gallons was carried in 32 miles of pipes, tunnel and redwood flume to Laguna Honda Reservoir, inside the City. First delivery was made in July 1862. Shortly after that, the company acquired rights at spring-fed Lake Merced in southwestern San Francisco near the ocean, and continued development of the Peninsula sources by building a dam to hold water in San Andreas Valley, west of Millbrae. San Mateo Creek water was impounded by Upper Crystal Springs Dam in 1878 and Lower Crystal Springs Dam in 1890. The latter was the final development of all practicable sources of water in the area south of the City.

In this water development by Spring Valley the most imposing structural achievement was the erecting of concrete Crystal Springs Dam, which still impounds up to 22½ billion gallons of water, serving as a terminal reservoir for Hetch Hetchy

Aqueduct. This arched, gravity-type dam, whose height has been increased twice, is 154 feet high and 600 feet long at the crest. The splendid manner in which this structure, located only a few hundred feet from the San Andreas Fault, survived the 1906 earthquake without the slightest damage, is testimony to its excellent design and construction under Spring Valley's renowned engineer, Herman Schussler.

As early as 1870 the City's engineers foresaw that water resources of the region would be inadequate for future population. A committee formed in that year had investigated Lake Tahoe, Clear Lake and Peninsula sources. There was a bill before the state legislature at that time to incorporate a company to bring water from "Lake Bigler" (Tahoe) by means of a tunnel through the Sierra Nevada mountain summit. The *Daily Transcript* of neighboring Oakland noted on March 9, 1870, that "none of the leading newspapers . . . have as yet commended [this solution]. The opposition only extends to the present bill, for every one [*sic*] perceives that some action is imperatively necessary." This newspaper then proposed Kings River as a source.

San Francisco's Board of Supervisors went ahead in favor of the Lake Tahoe Water Co. a year later, but Mayor Selby vetoed their $6,000,000 bond issue in April 1871, stating that the entire setup was improper and that he considered Tahoe public property. "Whatever action is taken, it should, in my opinion, be with a view to making the water supply a public enterprise, owned and controlled by the city," Selby said, as reported in the *Daily Transcript* of April 26, 1871.

In 1874 a second attempt was made to find a water supply outside what was offered by the private company. City and County of San Francisco officials sent newly-appointed Water Engineer T. R. Scowden on a scouting expedition, and he spent eight months making studies of possible sources at Blue Lakes, Clear Lake, Tahoe and other areas. As a result of this search, recommendation was made in 1875 that the City purchase Calaveras Valley in southern Alameda County as a future water source. But Spring Valley people wanted no rival and themselves bought this property before the City was able to act. Recurrent dry seasons and the inadequacy of water supply, demonstrated by numerous tragic fires, kept the problem always prominent in City politics. From 1867, when Spring Valley gained complete control, there was constant friction between the private water company and the consumer, the only breaks being brief periods of "armed neutrality." The company was always slow to increase its supply as population grew. Controversies were continually arising between Spring Valley and City officials as to rates and adequacy of service. The system, intrenchments and heaviest guns were on the side of "the monopoly," with disorganized and unsustained effort on the side of municipal ownership.

San Francisco made numerous attempts to gain ownership of its water supply. The Supervisors passed a resolution to purchase Spring Valley assets in 1875, and an $11,000,000 offer was made in 1877. No deal. It seems that purchase efforts were made principally in times of water shortages and dry seasons, being forgotten quickly when rainfall was abundant. A Supervisors' report of 1908 notes that "the indifference of the public generally is the true cause of our repeated failures" to buy out the water company. City officials in 1877 financed another year-long search throughout Northern California for a suitable water source. To the old locations they added Putah Creek, the San Joaquin, Rubicon River, Mokelumne River and others. The Blue Lakes scheme was recommended as most available, but negotiations were broken off for legal reasons.

A contemporary outside opinion appeared in Oakland's *Daily Transcript* of May 15, 1875, which stated that, for the "Spring Valley Crowd,"

> things appear to be quite well fixed. It can charge what it pleases now and it is not at all modest in piling up the cost. If the city, in an effort to become free, tries to procure a supply of water from some other source, the Spring Valley people come to the surface as owners of the lakes, rivers, or watersheds that the Engineer says would be most desirable. By adroit management this company will exact from the people an immense sum of money, no matter what course may be pursued by them. . . .

Three months later the same newspaper was asking whether or not the acts of the Spring Valley Water Co. and its officers would constitute a case of "criminal conspiracy" against the people of San

Pilarcitos was first of the Peninsula sources tapped by Spring Valley Water Company to obtain water for the City. This old sketch shows part of the wooden flume at outlet from the main tunnel. Final, and largest, of the Spring Valley dams impounding Peninsula water was Lower Crystal Springs Dam over San Mateo Creek. This 1890 view below, shows company officials on a trip of inspection during dam construction. Note champagne bottle on ground to the right. *(—Above, S.F. Public Library; below, Society of California Pioneers)*

Crystal Springs Dam south of San Francisco formed the largest of three reservoirs supplying water to the City. Construction, started in 1888, was of interlocking concrete blocks. It was undamaged by the big earthquake of 1906, although San Andreas Fault passes along the bed of the reservoir. This 1938 view shows excess water flowing over the spillway. *(—S.F. Water Dept.)*

Francisco. Yet the monopoly was to continue for three more decades, until the terrible disaster of 1906 finally enabled the voters to see through the deceptions and to act. In 1908 City Engineer Marsden Manson described the struggle as the "greatest of its kind in the water supply of American cities." At this time San Francisco was one of only two major cities in the United States relying on a private company for its water supply. Feeling was running strong for municipal ownership, preferably of a supply with Sierra Nevada sources to ensure dependability.

J. P. Dart, a civil engineer at Sonora, in 1882 suggested the Tuolumne River for the City's water supply. This stream rises along the crest of the Sierra Nevada in Yosemite National Park, flows westward in a tumbling rush through the mountains and foothills, and emerges in the great Central Valley to join the San Joaquin River flowing northwestward. The U.S. Geological Survey of 1899 published a report which stated that the river could furnish San Francisco with 250,000,000 gallons of water per day. This source of supply, however, was over 150 miles away. And the 150 miles was made up of about 50 miles across the Coast Range, another 50 across the low-lying San Joaquin Valley, and then 50 more miles across the Sierra foothills and well up into the mountains themselves! This was a tremendous challenge to be faced by a city with less than 350,000 population, supported by the few scattered communities around the Bay at that time. But it was a situation that had to be met, if the Bay region were to continue to grow.

Meanwhile, the Spring Valley Water Co. acquired extensive water rights, in addition to Calaveras Valley, in Alameda and Santa Clara counties, south and east of the Bay. In 1887 a pipeline was laid to carry water from the Alameda Creek area across and under the southern end of San Francisco Bay to connect with the Peninsula System at Crystal Springs. Further developments of the Alameda sources included sinking walls in the gravel beds of the Pleasanton area and construction of tiny Sunol Dam with a series of underground filtration galleries. This effort would reach its climax with the dedication of Calaveras Dam in 1925, but more of that later.

2

THE TWELVE-YEAR FIGHT

The City and County of San Francisco in 1900 adopted a new charter, containing a declaration for municipal ownership of public utilities, which spurred on the proper officials to take the first step toward acquisition of a water supply. The following year, during the administration of Mayor James Phelan, exhaustive surveys were made by City Engineer C. E. Grunsky, who reported on fourteen possible water sources. These included areas previously considered, plus the Yuba, Feather, Sacramento and Eel rivers, the Stanislaus, Bay Shore gravel beds and the plant and resources of the Bay Cities Water Co. On many of these, of course, there were prior established rights which would interfere with utilization of the natural flow by San Francisco. Even so, most of the proposals were represented by lobbyists, all of whom pushed mightily at City Hall and in the daily press. Noting opposition to the various water plans, the famous engineer, John Debo Galloway, wrote in the Berkeley *Civic Bulletin* for May 15, 1913, that the opposition "has come from various companies and individuals who seek to make profit out of the needs of the city by selling some other water supply. . . . Supplies ranging in size from 30,000,000 gallons per day to 500,000,000 gallons are freely offered for various sums depending upon how much money the promoters would like to have. . . ."

In 1902 the Board of Public Works had made its recommendation that the Hetch Hetchy Valley and Lake Eleanor on the Tuolumne River watershed were the best possible sources of water for San Francisco, both places lying within the northern reaches of Yosemite National Park, but several miles from the famous Yosemite Valley. The investigations had established the superiority of the Tuolumne River for the following reasons: purity of water, largest amount of water available, largest and finest reservoir site, freedom from conflicting legal claims, and electric power potential. The Tuolumne River, with its source in a perpetual glacier on 13,000-foot Mt. Lyell, drains 652 square miles of watershed in rugged granite mountains; 92% of the watershed is above 6000 feet elevation. Considering the partial pollution then existing in other sources and the rapid rate of pollution growth to which these others would be subjected in future years, the Tuolumne was absolutely superior. The City of San Francisco, besides talking about it, was one of the few cities that actually held out for quality in its water supply.

Following the decision to develop this watershed, steps were immediately taken toward acquisition of the necessary rights. Publicity was avoided as much as possible to prevent private parties and corporations, more mobile than the City, from forestalling the City's action. All applications for water and reservoir rights were made in the name of Mayor Phelan in his capacity as a private citizen. Phelan in 1903 assigned his interests to the City and County of San Francisco. Now the battle, started in 1901, hit full stride and continued unabated for the next ten years. The mountain development was named Hetch Hetchy Project, taking the name of a valley north of Yosemite.

This small, flat valley, three or four miles long and with steep granite walls, was to be the main storage reservoir. From prehistoric times local Indians had referred to it as "Hatchatchie," a Central Miwok Indian designation for a plant or grass, bearing edible seeds, which grew profusely in the

The upper reaches of the Hetch Hetchy watershed include areas of perpetual snow and a number of small glaciers. Investigation and survey parties covered all these and similar water sources of every major available stream in Northern and Central California. Left, a pack train carries supplies to a San Francisco survey party in eastern Tuolumne County. Below, San Francisco Supervisors stop at Hetch Hetchy Valley on a trip of inspection in 1910. *(–Left, Robert Cheminant; below, George Bartlett)*

Charlie Baird's stages out of Groveland served the Hetch Hetchy region on charter operation only. One of his outfits is shown here at Hog Ranch (later Mather) during the initial work period on the Hetch Hetchy Project, 1917. *(–Charlie Baird)*

San Francisco engineering teams surveyed all possible water sources before selecting Hetch Hetchy. This McCloud River survey party of 1909 included engineers Leslie Stocker, second left, and Max Bartell, far right. *(—Max Bartell)*

area of Hetch Hetchy Valley. Almost all the old-timers of Tuolumne County speak of the valley's unparalleled reputation as a breeding place for huge mosquitoes. It was in either 1903 or 1904 that a man named Stoddard developed the idea of making Hetch Hetchy into a big summer resort to rival Yosemite Valley. Charlie Baird operated a stagecoach from Hazel Green on the Coulterville-Yosemite Road to Hog Ranch (later Mather). The camp proprietors hired Baird to pack in tents, equipment and supplies from his headquarters at Groveland, and to carry passengers in as far as Hog Ranch. From there it was a four-hour trip on horses into the valley and some of the tourists had to be lifted off the animals and walked around when they reached the camp. Only about a hundred campers tried it out, Baird said, and the resort folded the same year.

The City's plan to build a reservoir within the boundaries of a national park precipitated a lengthy battle. By proposing to use federal lands the City placed itself in the position of being told what to do, tied and harrassed with restrictions of all sorts, penalized for several decades and having to fight for several years to nail down the rights it had entered upon. The war with Secretaries of the Interior over these rights was now on: they were to be denied twice by Hitchcock, granted by Garfield, modified by Ballinger, questioned by Fisher; finally the question would be appealed to Congress. Although conservation groups were the most vociferous of the opposition, this opposition was instigated primarily by corporate interests hostile to municipal ownership of public utilities or financially interested in rival water-supply schemes. Electric power corporations assisted the Spring Valley Water Co. with financial grants. Irrigation districts, users of Tuolumne waters for many years and stirred up by Spring Valley's Chief Engineer, were fearful that their supplies would be diminished.

The application filed in 1901, for permission to construct necessary dams and conduits and to use reservoir sites, was denied by Secretary of the Interior E. A. Hitchcock. The City continued its efforts to secure the necessary permission; persistent opposition came from three irrigation districts in San Joaquin Valley: the Waterford, the Modesto and the Turlock. Feeling ran high in the country, even though San Francisco did not propose to take from the Tuolumne River any of the natural flow of water. In January 1906, following receipt of a petition signed by 1200 landowners in the Turlock and Modesto districts, the Supervisors in the City formally abandoned their Hetch Hetchy plan on the grounds of its being illusory! Then came the earthquake in April, with serious impairment of Spring Valley Water's system, and renewed determination to by-pass this parasite. The Native Sons' *Grizzly Bear* magazine for February 1909 commented that it was lack of good principles by the water company which allowed the great fire to cost San Francisco $500,000,000 instead of a trifling sum.

Within just two months of the April tragedy eleven new proposals were received for furnishing water to the City. Of these, five were selected for further consideration: three on the North and Middle Forks of the American River, one on the

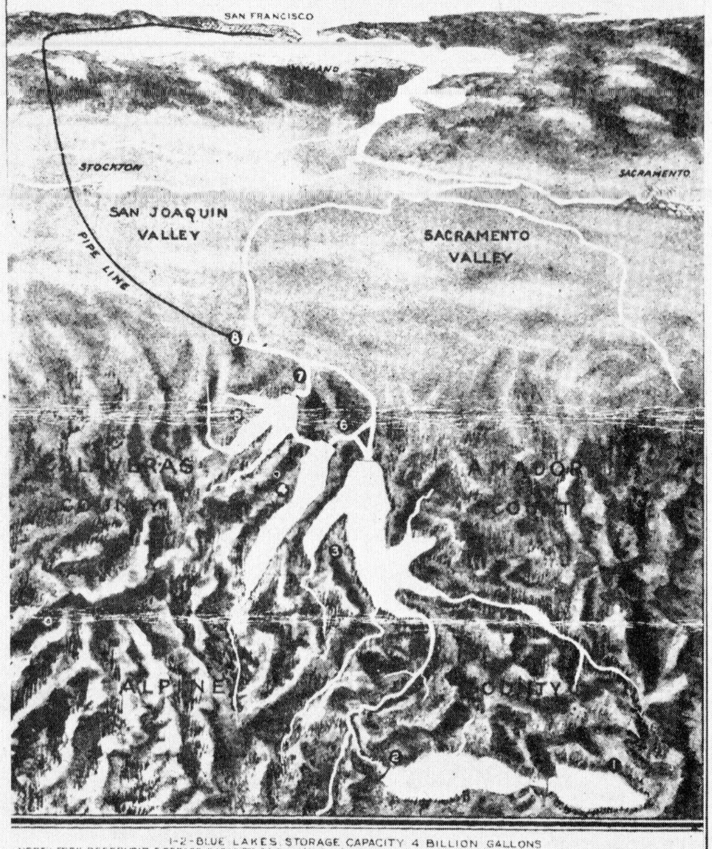

One of many schemes offered as a source of water for San Francisco was that shown above. The view looks westward from the summit of the Sierra Nevada, toward the City. The text is full of distortions and misinformation.

LET EVERYONE HELP TO SAVE THE FAMOUS HETCH-HETCHY VALLEY
AND
STOP THE COMMERCIAL DESTRUCTION WHICH THREATENS OUR NATIONAL PARKS

To the American Public:

The famous Hetch-Hetchy Valley, next to Yosemite the most wonderful and important feature of our Yosemite National Park, is again in danger of being destroyed. Year after year attacks have been made on this Park under the guise of development of natural resources. At the last regular session of Congress the most determined attack of all was made by the City of San Francisco to get possession of the Hetch-Hetchy Valley as a reservoir site, thus defrauding ninety millions of people for the sake of saving San Francisco dollars.

As soon as this scheme became manifest, public-spirited citizens all over the country poured a storm of protest on Congress. Before the session was over, the Park invaders saw that they were defeated and permitted the bill to die without bringing it to a vote, so as to be able to try again.

The bill has been re-introduced and will be urged at the coming session of Congress, which convenes in December. Let all those who believe that our great national wonderlands should be preserved unmarred as places of rest and recreation for the use of all the people, now enter their protests. Ask Congress to reject this destructive bill, and also urge that the present Park laws be so amended as to put an end to all such assaults on our system of National Parks.

Faithfully yours,

John Muir

November, 1909.

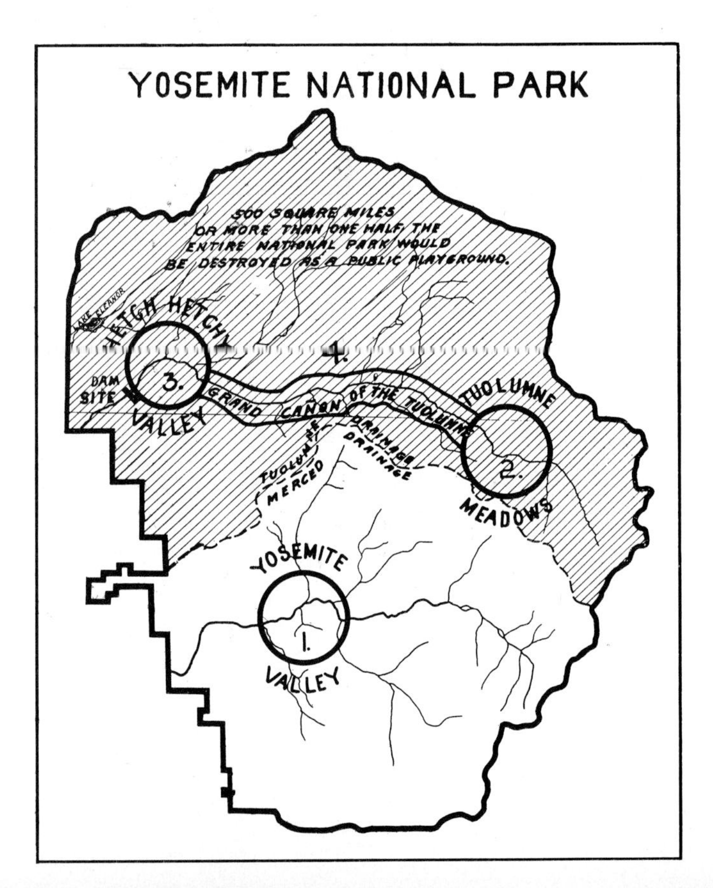

Right, cover page and map from John Muir's 1909 pamphlet opposing the Hetch Hetchy Project. The 32-page booklet contains a number of exaggerated statements, one of which appears on the map.

Marsden Manson, San Francisco's City Engineer under whom much of the research and planning for Hetch Hetchy was accomplished.

Mokelumne, and that of the Bay Cities Water Co., which would tap the South Fork of the American and North Fork of the Consumnes. While Mayor Schmitz's Board of Engineers was making a study of the proposals, a committee on water supply suddenly selected the Bay Cities Company's proposal to sell its system to the City for $10,500,000. All hell broke loose in the newspapers and City offices! Former Mayor Phelan's San Francisco *Bulletin* called it graft and bribery. Before a report could be made by City engineers on the properties, the graft scandal burst like a bombshell in the City. Mayor Schmitz was removed; sixteen of the Supervisors resigned and new men were appointed.

Secretary of the Interior James R. Garfield subsequently reopened Hetch Hetchy proceedings in San Francisco and in 1908 gave his approval, known as the "Garfield Permit." This allowed the City to go ahead by using Lake Eleanor as a reservoir site first and reserving Hetch Hetchy Valley for later development as needed. Rights of way for dams, reservoirs and aqueduct lines were also granted. It was a great victory and the people of San Francisco responded by voting almost unanimously a $600,000 bond issue. They followed in 1910 with another overwhelming vote for a bond issue of $45,000,000 to finance construction.

During the interim, however, those in the know up in Tuolumne County had not been idle. One William Ham Hall was aware that the City's plans called for a dam at Hetch Hetchy, but they had not officially included Lake Eleanor in the filings of 1901. Hall was thoroughly familiar with the region, having been State Engineer from 1878 to 1889 and U.S. Geological Survey Engineer, 1890-92. He knew what the filings meant, knew that development work had to be prosecuted to maintain them. So without touching Hetch Hetchy proper, in the interval between 1902 and 1906 he went into the open field and located all rights on Eleanor Creek and also on Cherry Creek (or River), which had been overlooked by City Engineer Grunsky in his filings. Hall and his associates, including world-renowned mining authority John Hays Hammond, bought up all lands in private ownership in the Lake Eleanor basin and in Cherry Valley, knowing that if the City went ahead with Hetch Hetchy, it would have to have their holdings.

So now, when Secretary Garfield gave the government's permit with the stipulation that Eleanor should be developed first, everything involved was in the hands of Hall's Tuolumne Water & Supply Co. The City had to deal with Hall, finally paying $400,000 for the Eleanor holdings. When it was further discovered that this would not be adequate for the City's needs, they had to go again to Hall and pay $652,000 for the adjacent Cherry group of rights – a total of a million dollars outlay caused by delays.

In order that the City might proceed, a joint resolution had been introduced in Congress in 1908, providing for an exchange of lands between San Francisco and the federal government, as a provision of the Garfield Permit. San Francisco had title to various lands outside Hetch Hetchy Valley, to be exchanged acre for acre for government land in the valley proper. When hearings opened in Washington in December 1908, it was found that practically the entire country had entered the debate! There was a deluge of protesting letters, resolutions, and attacks in the press against allowing national park land to be used for water conservation. Spring Valley Water's attorney appeared without announcement before the committee, presenting legal objections and opposing this exchange. The water company charged that the City's real object was to use Hetch Hetchy as a club to force it to sell to the City. Back in San Francisco a storm was raging over the private utility's opposition at Washington. Proponents of municipal ownership and Hetch Hetchy enthusiasts became more determined than ever. City officials and businessmen fought back with letters, magazine articles and newspaper features of their own, referring to Spring Valley as "the impudent monopoly."

Writing in the *California Weekly* of June 18, 1909, City Engineer Marsden Manson objected to falsehoods in the writings of John Muir and his followers in their attacks on the City. There were accusations that San Francisco was attempting to destroy the wonders of nature in the interest of saving money. This charge was answered by James Phelan, devoted ex-Mayor, later United States Senator and diplomat, in an impassioned paragraph in the same weekly:

> It is not "cheapness" but abundance and purity and reliability that move the people of the Bay of San Francisco to provide for themselves and their posterity in time; but if it were partly a question of reasonable cost, I think that the people of the country would consider it no small element in the petition of a city that has just been destroyed by fire, due to a lack of water, and which is now using her last available penny, under the limits of taxation, to re-create a municipal plant, and to worthily serve the nation as its western gate. She should not be falsely condemned for the lack of esthetic appreciation of the beauties of nature, nor reproached for thrift.

Engineer John D. Galloway came forward to defend the City and to point out that the real opposition, those who financed the battle and encouraged opposition to the City's acquisition of Hetch Hetchy water, were the people who had approached the graft-ridden regime of Mayor Schmitz through Abe Ruef, offering over a million-dollar payoff if the City dropped Hetch Hetchy and bought the Bay Cities Water scheme. Galloway said also that the violent storm of opposition was mainly inspired by the water company, with nature lovers and "some members of the Sierra Club" making the most absurd statements and showing "utter ignorance of conditions" of obtaining water elsewhere.

There seems to be no doubt that John Muir was sincere in his struggle against the use of Hetch Hetchy Valley as a reservoir. And his impassioned writings had nationwide repercussions and attracted wide support. In a 1909 pamphlet Muir wrote: "Dam Hetch-Hetchy! As well dam for water-tanks the people's cathedrals and churches, for no holier temple has ever been consecrated by the heart of man." His campaign to build sentiment against the plan in spite of the fact that most Californians were in favor of using Hetch Hetchy, drew a rejoinder from Benjamin Ide Wheeler, President of the University of California. Wheeler charged that "Many of the best people in the [Sierra] club . . . do not agree with Mr. Muir."

Early in 1910 the Mayor and Board of Supervisors were surprised to receive a letter from Richard A. Ballinger, Taft's Secretary of the Interior, asking them to show cause why the Hetch Hetchy Valley and reservoir site should not be eliminated from the Garfield Permit of 1908. Many people in California considered the Taft administration's opposition as a grievance and there were political implications. At the direction of President Taft a board of three Army engineers was assigned in 1911 to investigate and report on the subject of water supply for San Francisco as bearing particularly upon the need of the sources of the Tuolumne. The board consisted of Colonel John Biddle, Lt. Colonel Harry Taylor, and Major Spencer Cosby, and they examined exhaustively all alternate sources of supply which had been suggested as available for the City's use. Data were gathered by teams of engineers and surveyors; the Army board went to examine Hetch Hetchy with two platoons of cavalry and a government pack train of 25 animals; and the City called in John Freeman, an engineer of national repute, to make an exhaustive examination and come up with a proposal.

The late Max J. Bartell, who worked almost continuously for the City as an engineer from 1908, said that Freeman came and sat three weeks in his office and absorbed all the data. Then he set

Why Hetch Hetchy Is Necessary

The Water Supply: Warning!

The water consumption in San Francisco now exceeds the safe, dependable supply available for distribution. Until the city or the company can increase the development of sources now owned and install more aqueducts to San Francisco, extreme care must be exercised in the use of water.

Or the Supply Will Fail. Stop All Waste; Stop Hosing Steps and Sidewalks With Water. Please Prevent All Unnecessary Use of Water. We Earnestly Ask for Your Co-Operation in Maintaining the Supply.

SPRING VALLEY WATER CO.

This notice from Spring Valley Water Company was reproduced in the San Francisco *Examiner,* December 2, 1913, to emphasize the necessity for congressional approval of the City's application. Below, a party of San Francisco Supervisors visits Hetch Hetchy Valley on an inspection tour in 1910, led by packer-guide George Bartlett, the man on the right. *(–George Bartlett)*

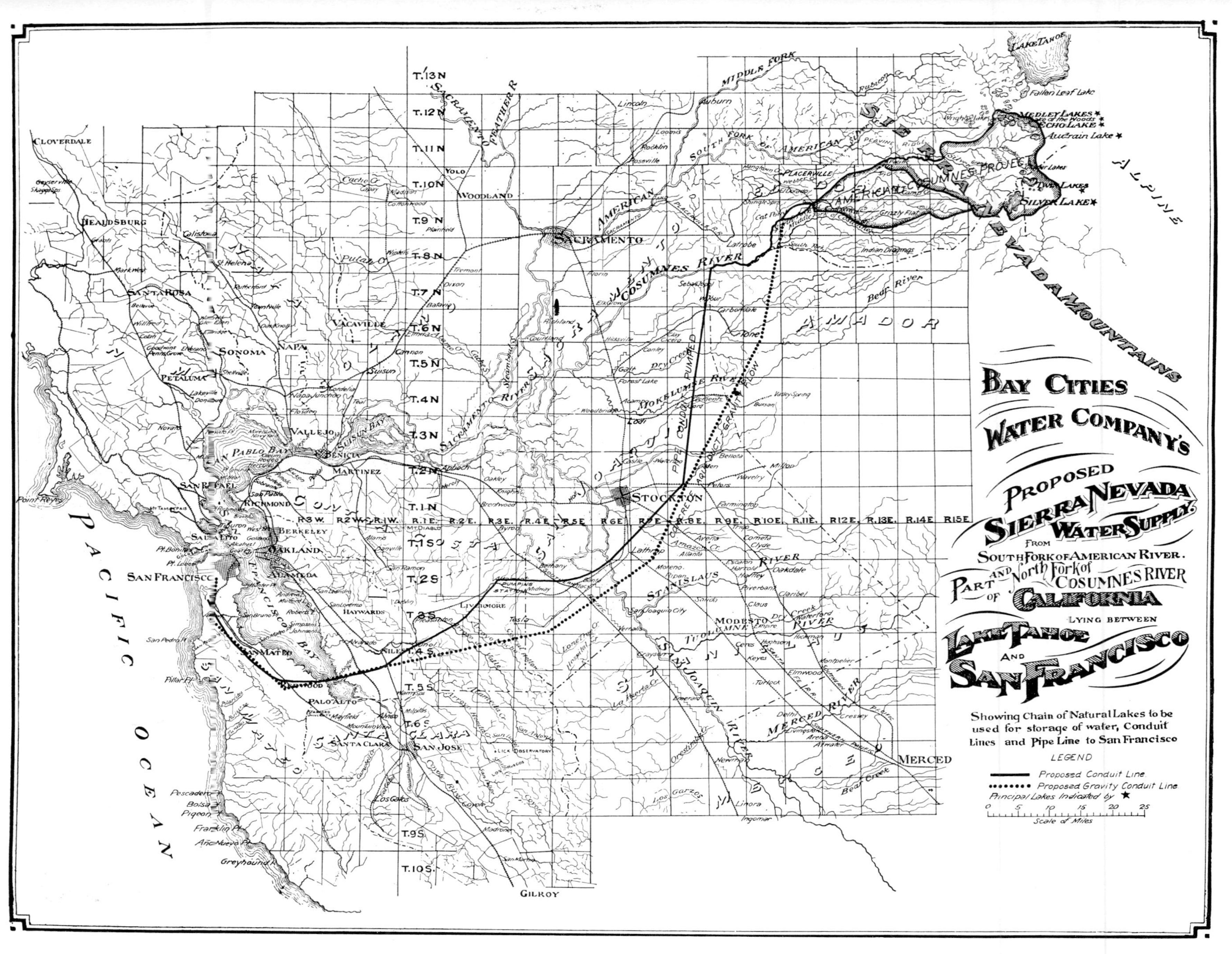

Proposal of the Bay Cities Water Company to supply water from the South Fork of the American and North Fork of the Consumnes River. It had been accepted by the graft-ridden regime of Mayor Schmitz.

about elbowing Marsden Manson out of the City Engineer's job, despite the fact that Manson had gone to Washington at his own expense in 1908 and held the status quo when the crooked Supervisors wanted to give away Hetch Hetchy. Freeman didn't get the job, but as a result of his investigation there was immediately published a large book of facts and proposals known as the "Freeman Report," ready in 1912 and presented with other reports and testimony to the Army board toward the end of its two years' investigation. As was to be expected, Spring Valley was on hand with its own massive study, a large bound book being mainly an answer to Freeman's proposals.

While the Army people were compiling their report, a new Secretary of the Interior, Walter L. Fisher, came to attend a conference in Yosemite. San Francisco had a new City Engineer, Michael Maurice O'Shaughnessy, and the latter was anxious to meet Fisher and present his case for the City's plans. He wired George Bartlett, a mountain packer and guide then in the City's employ, to meet him at Hamilton Station (Buck Meadows) with Bartlett's "12 Cadillac," which had the first electric lights and a self-starter. They drove east on dusty Big Oak Flat Road and up through the redwoods to Crane Flat, where the national park's official Overland car was to meet them, for no private autos were allowed into the park until 1912.

A ranger at the gate phoned into Yosemite and found that the car had left nearly two hours before and had broken down. He refused to let them through. But as soon as he turned his back, they headed for the valley. At the meeting in the Sentinel Hotel, Head Ranger Townley said "I've got to arrest you, George; come over to the clubhouse." Then he let them "escape" by way of Bagby and Coulterville. Many years later Bartlett told the author that he had the impression "O'Shaughnessy did a lot of good" that day with Secretary Fisher.

The Army engineers' report was made to Fisher in Washington on February 19, 1913. It recommended use of Hetch Hetchy Valley and the Tuolumne supply as being best suited for San Francisco's purposes. The statements of its Chairman, Colonel Biddle, formed the most exhaustive, detailed and apparently accurate testimony given upon available water supplies in this case. Secretary Fisher gave it as his opinion that Congress alone had power to grant the privileges sought by San Francisco. And the City was now more than willing to go to Congress and end once and for all the possibility that it would be subject to the whims of successive administrations in Washington. Congress was asked in 1913 for an outright grant of the desired privileges.

A recent view far above Hetch Hetchy, showing the Tuolumne River and its polished-granite, snow-covered watershed, which offered a reliable source of pure drinking water to the people of San Francisco. *(—S.F. Public Utilities Commission)*

There followed an extended and heated battle in Congress, with the private water company and its spokesmen making inflammatory claims and charges. The City's battle was now in the hands

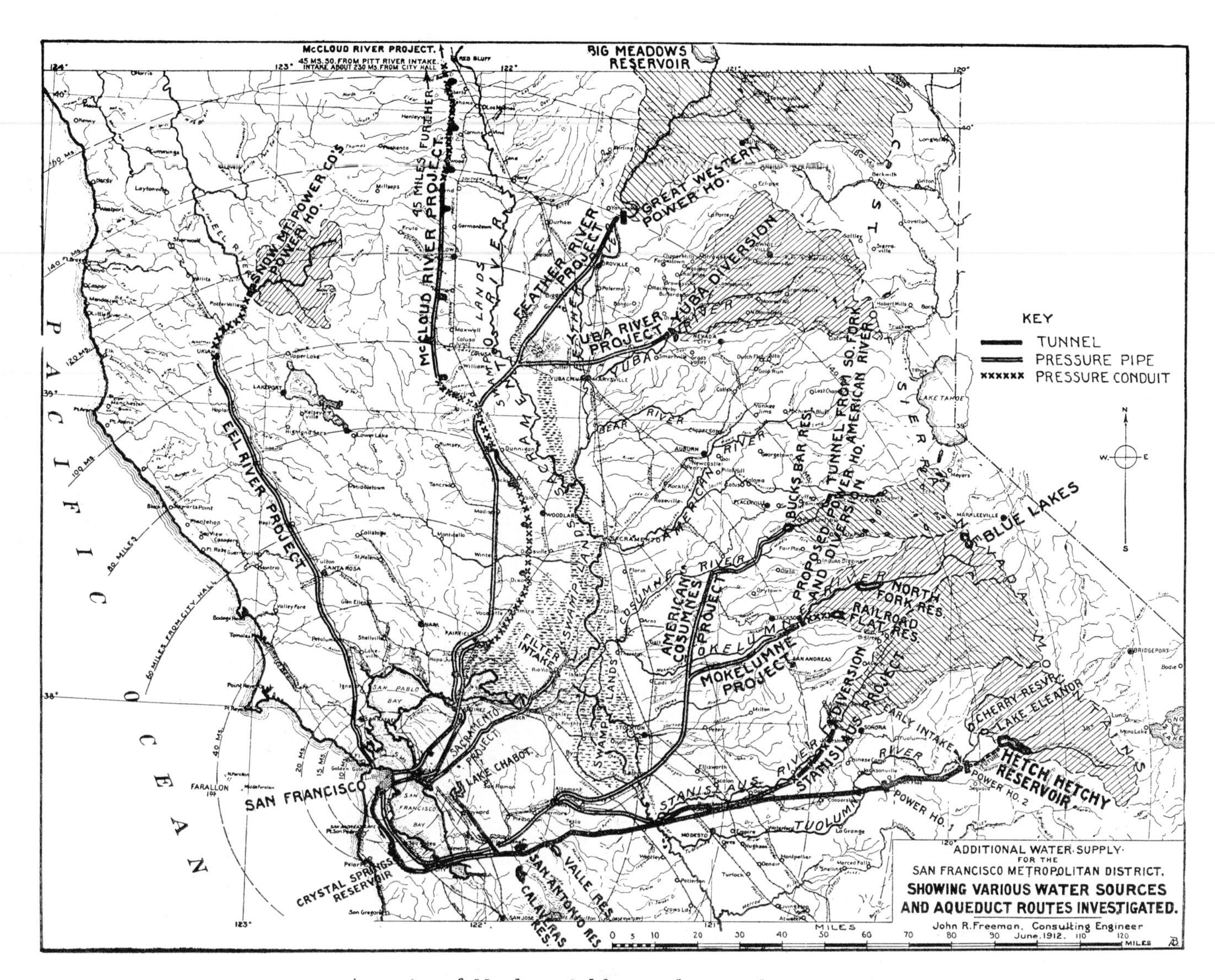

A portion of Northern California, showing the extent of investigations made by the City and County of San Francisco in its search for a source of water. This map and that of the Bay Cities Water Company proposal were published in the "Freeman Report" of 1912.

of Michael O'Shaughnessy; the previous City Engineer, Marsden Manson, had devoted twelve years almost exclusively to the Hetch Hetchy battle and it had worn down his health. In Washington there was a new President, Woodrow Wilson, and a new Secretary of the Interior, Franklin K. Lane, former San Francisco City Attorney.

John Muir and the Sierra Club attacked through letters, pamphlets and circulars, the club now basing its arguments on the claim that there was an adequate water supply for San Francisco without the need to "destroy" Hetch Hetchy Valley. This source was Mokelumne River to the north; they maintained that it would be sufficient, with water from Lake Eleanor and Cherry Valley and existing Spring Valley sources. Taggart Aston, from the Sierra Club, wired Washington in July 1913, supporting the claim of the Sierra Blue Lakes Water & Power Co. that the Mokelumne could supply the City's water needs. Aston also stated that the City's plea of a "water shortage" was not made in good faith, since there was plenty of water in Spring Valley's reservoirs. What a misrepresentation this latter statement was will be shown in the

Shown above is a 1913 view of Hetch Hetchy Valley, looking upstream. The view at left, five years later, shows the area just prior to the start of dam construction. Vegetation has been cleared from the valley floor. Kolano Rock, to the right, rises 1800 feet above the valley, while beyond is the rounded top of Rancheria Mountain. *(–Both, City & County of S.F.)*

subsequent story of Hetch Hetchy. The further allegation that City Engineer O'Shaughnessy suppressed a report on Mokelumne supplies is a prominent part of a recent Sierra Club book on John Muir and Yosemite. Except for the "suppressed Bartel-Manson [*sic*] report," it is claimed, the City might not have won its victory before Congress.

In 1912 Max Bartell prepared, among others, a report on the water of Mokelumne River. He concluded that there would be much litigation with private owners and that San Francisco would ultimately need more water than was available from the Mokelumne River. On the Sierra Club's copy of this report there were penciled notations, barely legible, which stated that this supply, with Eleanor, would be sufficient. Interviewed in 1966 and 1967, Bartell denied making such notations on his report because he never considered such an alternative. It was his opinion that any such notations were made after a copy of his report was removed from the City Engineer's office in 1913. Indeed, a copy of this same Bartell report of 1912 is at the Water Resources Center Archives, University of California at Berkeley. There are no notations or changes in this copy, the cover of which is stamped "OFFICE COPY / M. J. BARTELL" and an inked notation "Property of C. E. Healy" (Clyde Healy, Assistant Engineer).

The Hetch Hetchy Grant, known as the Raker Act, was passed by both houses of Congress and signed into law on December 19, 1913, by President Wilson, who made the following written comment upon the plan:

> . . . It seemed to serve the pressing public needs of the region concerned better than they could be served in any other way, and yet did not impair the usefulness or materially detract from the beauty of the public domain.

The Raker Act, named for Congressman John E. Raker of California, granted to San Francisco rights of way and use of public lands in the areas concerned for the purpose of constructing, operating and maintaining reservoirs, dams, conduits and other structures necessary or incidental to development and use of water and power. This was not, however, a free gift to the City of San Francisco, because the City had a number of obligations under the act:

1. Construct miles of scenic roads and trails in Yosemite National Park and donate them to the United States;
2. Enforce sanitary regulations within the watershed area;
3. Recognize prior rights of Turlock and Modesto Irrigation Districts to receive specified amounts of water for their storage and use;
4. Complete the dam at Hetch Hetchy as rapidly as possible;
5. Develop electric power for municipal and commercial use;
6. Not divert beyond San Joaquin Valley any water not required for its own domestic or municipal purposes;
7. Not sell or give Hetch Hetchy water or power to a private person or corporation for resale;
8. Pay an annual rental starting at $15,000 and rising to $30,000 after twenty years.

The Raker Act is an excellent demonstration of the "conservation for use" policy, whereby a great advantage was gained for millions of people and very slight damage done.

The City and County of San Francisco was still engaged in one of history's greatest construction jobs – rebuilding a city destroyed by earthquake and fire in 1906 – when work on Hetch Hetchy began in earnest in 1914. But this was a city loaded with engineering talent of the highest order, with City engineers and private consultants trained and tested in many of the West Coast's great projects. Some of the finest engineering brains of the age signed up with San Francisco because they liked the concept and they respected "The Chief," Mike O'Shaughnessy. And the inducement which drew this outstanding man to the job was the guarantee by Mayor Rolph of a free hand and ample support.

Numerous obstacles besides mountains and engineering problems had to be overcome. There was a 75% rise in prices from 1913 to the end of World War I, with a shortage of manpower at the same time, and there were attempts at political interference and foot-dragging on appropriations. Many prophesied failure on account of the gigantic nature of the project. But San Francisco flaunted the title "The City That Knows How," and all threw themselves wholeheartedly into the job.

3

TUOLUMNE COUNTY BACKGROUND

Center of the forested mountains of eastern California, Tuolumne County nestles between perpetual snows of the Sierra peaks and warm, flat agricultural lands of the San Joaquin Valley, due east of San Francisco. Great mountain streams flow out of the snows and glaciers, roaring westward in their downhill tumble to the San Joaquin River. Torrents of water habitually flowed out of the mountains in late springtime, usually May and June, when the melting of winter's snows accelerated under ever warmer sunshine; vast flooding of the lowlands was a rule of life. The rivers are now largely tamed by dams and some of the annual excitement is gone, but the county's gorgeous mountain scenery remains. Ridges covered with pine and cedar march grandly eastward from the foothills as far as the eye can see. Included in Stanislaus National Forest are 1130 square miles of the county, and a vast area at its eastern extremity is a part of Yosemite National Park.

This particular area was one of the great sources of gold during the days of '49. The first white men to come in any numbers came as gold-seekers, all bent on making fortunes and scurrying back to civilization. Every creek, river and gulch was raided by treasure-seekers, first with pan, then rocker, and finally sluice. Gold poured out by the bucketful. Early in the 1850s quartz veins were first worked; derricks, rockers, sluices and hydraulics gave way to hoists and mills. In due time supplies of water were brought in mining ditches and flumes, for water was absolutely necessary in extraction of the precious metal, and water is the greatest of the treasures of Tuolumne County.

Running east to west near the southern edge of the county, through meadow and valley and chasm, is the mighty Tuolumne River. It is the region along this river which will be the chief concern of this book. The great river and its branches, or "forks," will play the major role, with wild encircling mountains added, and that special breed of people born and raised "south of the river." Dubbing themselves "Poison Oakers," these natives still consider themselves a type quite removed from residents of the population centers around Sonora, the county seat rather far to the north. It was to Poison Oaker country that San Francisco came to seek its water. What was it like before the "city slickers" arrived?

The first discovery of gold in the county was made on Woods Creek, which flows west and south from Sonora and Jamestown to join the Tuolumne at Jacksonville. Gold mining was still big business all along this creek in the first years of the 20th century. A mile or so north of Jacksonville was the famous Eagle-Shawmut, largest mine in the county with a 2300-foot shaft and a hundred-stamp mill; 250 men worked there. South of the Eagle-Shawmut was the Republican group of mines, then the Clio, Joy and others, ending at the Oliver-Harriman straddling the Tuolumne. Continuing south from the pioneer town of Jacksonville were the North Star, the Black Warrior and several more in the canyons of both the Tuolumne River and diverging Moccasin Creek.

From the Moccasin there began a long, backbreaking climb to the mines up above, clustered around Big Oak Flat and Garrotte, as Groveland was known until 1875. These diggings were only a few miles away as the crow flew, but were 2000 feet higher in elevation. The Mount Jefferson and Cosmopolite mines were producing at Groveland,

Big Oak Flat Road was a toll road into the Hetch Hetchy region and Yosemite until 1915. This 1908 view shows the Groveland-Crockers Station mail stage stopped at the toll-house – the covered bridge over South Fork of the Tuolumne, about 15 miles east of Groveland. Francis Elwell was the driver and Lewis Elwell, tollkeeper. It was here that the famous Cliff House resort was located in recent times. Transportation into Tuolumne County had been considerably improved with coming of the Sierra Railway of California in 1897. Below is shown engine No. 2, just arrived with the train from Oakdale, at the temporary end of track, Chinese Station. Stages connected for all points, the nearest being that for Groveland, William Lawron driver. *(–Left, Tuolumne County Museum; below, Roy D. Graves)*

Crockers Station was the last stop on Big Oak Flat Road on the way to Yosemite. The view above, taken in the 1890s, shows the hotel at center, barn in left distance. Along the same road a number of "attractions" had been marked by an early staging company in an effort to lure travelers away from the Coulterville route. One was the famous "Hangman's Tree," across the road from "Bret Harte Cabin" at Second Garrotte. The last remnants were destroyed by a fire in 1970. *(—Gaytes Photos, Bancroft Library)*

Travel along Big Oak Flat Road was by horsepower or on foot in the '90s. The top photo shows a horse and buggy stopped at Jordan's Cash Store in the village of Big Oak Flat, two miles west of Groveland. Somewhat farther west a freight wagon, drawn by eight mules and carrying a cable spool for the Eagle-Shawmut Mine, pauses at Chinese Camp. *(—Gaytes Photos, Bancroft Library)*

About 12 miles out of Groveland the stage bound for Crockers and Yosemite skirts the edge of the cliff above the South Fork of the Tuolumne River, near Colfax Springs. The Hetch Hetchy Railroad was later put on another shelf below the road here. *(—Selina Lumsden)*

Groveland and Big Oak Flat were settlements of the Gold Rush era that managed to survive into modern times. Gold mining went on actively in the area into the 1930s. This view shows the headworks of Mt. Jefferson Mine just outside Groveland in 1898. *(—Catherine Cobden Haight)*

while at near-by Big Oak Flat were the Mack, Longfellow, Nonpariel and Red Jacket. As in all mining, it was sometimes good, sometimes bad, but hundreds of men were employed in that area in the decade before the Hetch Hetchy Project began.

Vast supplies of water needed for these and numerous smaller mine operations along Big Oak Flat Road, on Big Creek, Rattlesnake Creek, in Grizzly Gulch and other "south of the river" areas were carried in the famous Golden Rock Ditch. Varying in length over the years from 38 to 45 miles, this combination of flume and ditch took water from the South Fork of the Tuolumne at Hardin Flat, held it at the higher levels and made distribution along the way. Orchards and ranches were served as well as mines, since this section of the county had made a partial transition to farming and ranching. Golden Rock Ditch became part of the Tuolumne River Power Co. in 1905, then the Yosemite Power Co. in 1911, with vast plans for reservoirs at Hardin, Ackerson, Poopenaut Valley and other places, and great amounts of electric power to be produced.

Indians of the Miwok family were very much a part of the Groveland scene until the turn of the century. The Indian village just outside town was popular with the boys and young men, who went down to gamble and to watch the feasting and dancing at annual councils. Groveland and Big Oak Flat were the only settlements of any size in the area, although only a couple of miles apart and separated by a ridge. Western characters abounded: the last of the original '49ers lounged on sidewalk benches; cowboys trailed constantly through the towns; the roads were traveled by packers and guides with their animals. During the 1900-1915 period a daily stagecoach with four tired horses arrived from the west late at night, bringing mail and passengers from faraway Chinese Station on the then-new Sierra Railway; it left town at 4:00 a.m. on the return trip. A couple of times a week another stage tackled the mountains to the east with mail, supplies and a very few passengers for Crockers Station at the edge of Yosemite. Stagecoaches for the national park left frequently on charter, because tourists were not allowed to take autos into Yosemite.

Freight teams of ten mules arrived and departed constantly with machinery and supplies for the mines, with food and other necessities. Although the era of the autostage was imminent and the motor truck would soon appear on the scene, these were delayed by road conditions in this country. Only horse-drawn vehicles could possibly get through that axle-deep mud in winter. And the awful Priest Grade coming up from the west, although hard enough on horses and mules, was a tremendous barrier to early motor vehicles. As the county chamber of commerce noted in a 1909 book, "What is needed for Big Oak Flat and Groveland and the rich country stretching away for miles in all directions . . . is a railroad, and they are certain in a short time to have one. It's a case of absolute necessity."

The Yosemite Short Line Railroad of a few years earlier would have served the purpose.

It was a dirt road to and through town until the 1920s. Only horses could travel beyond Groveland in the winter mud, and the axle-deep dust of summer covered vehicles and riders alike. Top view shows Groveland's "main street" – the Big Oak Flat Road – and the new bridge of 1911. The automobile had arrived in numbers by 1916, the center photo showing "downtown" Groveland with the hotel at center and the Iron Door Saloon to the right. At the bottom is George Brown's team, Henry Henniker driver, with a load of mine timbers a couple of miles east of Second Garrotte in 1905. *(–Top, Grace Stanley McCaslin; center, Irene Magee; bottom, Bancroft Library)*

A 1914 view of Hetch Hetchy Valley and the Tuolumne River. *(—Photo by Ernie Beck)*

On the road to Hetch Hetchy on a warm summer's day in the '90s. The bridge crosses the Middle Fork of the Tuolumne, near the later location of the Oakland and San Jose recreation camp. *(—Gaytes Photo, Bancroft Library)*

In 1912 the people of Big Oak Flat and Groveland, assisted by merchants and mining companies, set out to build a decent road up Priest Hill, since the county could not or would not tackle the job. Here a group of volunteers takes a break from picking and shoveling to have their picture snapped. Seated left is Charlie Baird, who provided free stagecoach transportation from town; others, left to right, are Ed Watson, Charlie Hall, Roscoe Gray (kneeling), Jake Laverone, Eugene McCarthy, Charlie White, (not identified), Sy Goldsworthy, Jr., Bob Nielson, Charlie Harper, Jack White and Sam Blagg. *(—Selina Lumsden)*

East of Groveland wintertime travel was often made impossible by heavy snows; without horses, there would have been no communication whatever. Gray's wagon and team are shown near Jones Station beyond Middle Fork just about the time the Hetch Hetchy Railroad was building into the area.

Carrying supplies up to Groveland, George Brown's freight wagons and 14-mule team stop for a rest at Slade Gulch between Priest's and Big Oak Flat in 1914.

Shown left is the Eagle-Shawmut Mill with incline descending the hill from the mining area. Largest mining operation in Tuolumne County, this was located on Woods Creek two miles north of Jacksonville. The 100 stamps made a roar that could be heard for miles. *(—Both, Tuolumne County Museum)*

On the way to Hetch Hetchy, Jacksonville was on the Tuolumne River a few miles out of Chinese Camp. This 1916 view shows horse-drawn and Model T delivery wagons and a banner urging "Ice Cold Valley Brew." Sheafe's Hotel, to left, was dismantled by Charlie Baird and rebuilt as Baird's Hotel in Groveland. The photo at left shows the Groveland Hotel and snow-covered Big Oak Flat Road in winter, 1917, looking west. Bottom, famous Priest's Hotel at the top of infamous Priest Grade was a favorite stopping place for travelers headed for Yosemite. The hotel and annex in foreground were destroyed by fire in 1926. *(—Top, City & County of S.F.; bottom, Tuolumne County Museum)*

The Yosemite Short Line Railroad, headed for timber and Yosemite Valley, got as far as Jacksonville in 1906, started to bridge the Tuolumne River and gave up. The bridge piers were washed away one by one, but these few were still there when everything was leveled for a new Don Pedro Reservoir in 1970. Photo at right, shows Big Oak Flat Road at entrance to the then-new Oakland Recreation Camp, 1921. "Overnight Camp for Auto Parties" on the Middle Fork was 15 miles east of Groveland.

Started in 1905 from Quartz Junction, near Jamestown, and backed by Sierra Railway interests, this narrow-gauge line was intended to pass just north of Groveland on its way to Yosemite, tapping rich timber tracts owned by the backers. But after it reached the river crossing at Jacksonville, very little additional work was done; the project was given up and the two Porter tank locomotives departed. In a few years, however, a railroad was to come to the mountains of southern Tuolumne County as a result of the work on the Hetch Hetchy Project. And the Poison Oakers were ready, having followed San Francisco's battle in Groveland's weekly *Tuolumne Prospector*. They all seemed to favor the plans of "The City": they needed the activity; life was slowing south of the river.

Although still the county's biggest industry, mining had decreased in the first years of the new century. Power from the Tuolumne Electric Co., with a plant on the river north of Groveland, had arrived in 1907 and rejuvenated the mines for a while, but it was subject to frequent breakdowns. Farming and ranching in the southern parts of the county were important but on the verge of extensive cutbacks. Water service from Golden Rock Ditch couldn't be counted on and the district was again going arid. Serious logging and lumbering had started in earnest along the new line of the Sierra Railway, but in Poison Oaker territory it still awaited cheap and reliable transportation.

Residence in that district, before the arrival of the "Hetch Hetchy gang," was definitely losing its attractiveness. There was constant complaint in small weekly newspapers about road conditions: impassable mud in winter and unbearable dust in summer! "Up on the hill" at Groveland and Big Oak Flat, people were tired of being ignored and neglected by the rest of Tuolumne County – this was a constant refrain. Note this line in the *Prospector* for July 28, 1906: "The telephone at Groveland has been out of order for the past few days and consequently Groveland has been dead to the world."

This was all due to change rather suddenly, as San Francisco came into the area in force to tackle its Hetch Hetchy Project. Groveland was destined in a few years to be the center of more activity than it experienced in all its history, before or since, all compressed into six or seven years of full-speed go-Go-GO!

San Francisco was still cleaning up after the 1906 earthquake and fire when the City undertook to build its Hetch Hetchy water and power project. Derrick loads rubble on a truck at Fulton Street and City Hall Avenue, June 1917. In the background is St. Boniface Church.

San Francisco sent this survey party to Hetch Hetchy in 1912 to locate section corners in the proposed project area. Men of that group, on their way home at summer's end, wait for the westbound train at Sierra Railway's Chinese Station. Emile Muheim and Billy Eggert to left. *(–Anna Muheim)*

4
STARTING THE BIG JOB

The Hetch Hetchy Project was a tremendous undertaking. The fact that it was tackled by a city alone, without state or federal funds, is even more tremendous by modern standards. Further, the work in the mountains was complicated by the inaccessibility of that area. In a country difficult for mountain climbers, with only a few spots offering even enough feed for horses, it was proposed to build huge dams and electric power plants, transport all manner of machinery, equipment and supplies, and feed and care for thousands of workmen. As for Big Oak Flat Road, which served this region and eventually reached Yosemite, former Hetch Hetchy employee "Red" Fent says it was laid out by following a jackass into the mountains, using the animal's hind end for a transit! But now no area could be considered inaccessible since electrically driven drills had been developed to dig into granite and dynamite was available as the moving force.

Plans for this job had been drawn up by two City Engineers, Carl E. Grunsky and Marsden Manson. Under Grunsky, surveys had been made leading to selection of the Tuolumne and acquisition of various rights. Manson devoted nearly all his energies to furthering the Hetch Hetchy scheme. Max Bartell and Leslie W. Stocker, who had started their work as engineers for the City under Manson and continued all through the O'Shaughnessy regime and beyond, both stated many years later that Manson was truly the "father of Hetch Hetchy Project." He had a party out surveying boundaries in 1908 under Drenzy Jones, former Tuolumne County Surveyor. With him were two assistant engineers from the City, Les Stocker and Louis Mercado. It was extremely difficult work, running out land boundaries at Hetch Hetchy, Hog Ranch, Tiltill Valley and other places – up and down near-perpendicular cliffs and straight across roaring streams. They'd been given a month for the job, but it was only two-thirds completed when winter forced the party out of the mountains. In 1909 Jones directed several parties in the surveys, the men going into Hetch Hetchy part-way on the narrow-gauge Hetch Hetchy & Yosemite Valley logging railway.

Surveying continued up into 1912. Many tentative aqueduct lines were run out; surveys were completed for roads, trails and tunnels. Thus, by the time investigations were started by consultant John R. Freeman in 1912, the City had accurate data to go on in laying out the work on the ground. At this point Mayor Rolph had asked O'Shaughnessy, a man with superb qualifications, to take over direction of the project. A graduate of Royal University in Dublin (1884), the new City Engineer had behind him over 25 years' experience in California and Hawaii, including San Francisco's Midwinter Fair of 1894 where he was chief engineer. When tapped by Rolph he was 48 years of age, serving as chief engineer for the Southern California Mountain Water Co., and had recently designed and built its successful Morena Dam near San Diego. It was generally understood that Mike O'Shaughnessy took a cut in pay when he accepted the San Francisco job. But the "fringe benefits" were attractive: he would answer only to the Mayor, just as long as the job progressed; there would be no interference from the Board of Supervisors or any other politicians. And this project was just the sort of challenge O'Shaugh-

Autos were scarce when Hetch Hetchy Project work was started in Tuolumne County. Above, one can be seen tucked into Baird's livery stable at Groveland in 1914, while the stagecoach sits out in the rain. At Chinese Station in 1916, below, a large touring car stands beside the road to Yosemite in this view from the railroad station. Hotel Sims is behind the tree to left. *(—Above, Ernie Beck; below, F. S. Rolandi, Jr.)*

A light blanket of snow touches the eastern end of Groveland in February 1917 as work on the Hetch Hetchy Project gets under way in earnest. Just to right of center the large building with covered, two-story porches is the newly completed San Francisco headquarters. To the left of it, in the flat area along Garrotte Creek, would be laid out the railroad shops and yards. The voters of Groveland, early in 1973, approved purchase of this area for their town park. *(—City of S.F.)*

nessy liked; here he could work on his theory that the "only final solution of water supplies in California is by construction of large dams on the high levels of the mountains to store floods."

O'Shaughnessy gave a fighting speech at Oakland in 1916, when inviting East Bay cities to become partners in Hetch Hetchy. After thanking Senator Perkins of Oakland and Congressman Joseph Knowland, both of whom had helped in the Raker battle, the fiery San Francisco engineer concluded with these words, reported in Knowland's Oakland *Tribune* (October 7): "Hetch Hetchy is ours. We'll do what we wish with it. We'll build where we wish first. It is ours and no one will take it away from us. Our men are there; our road is there. We have begun work, and we will stay there until we complete our work and we have Hetch Hetchy water in San Francisco."

Actual construction work began as soon as the congressional grant was obtained. However, although the people had voted a $45,000,000 bond issue in 1910, a considerable amount of difficulty was experienced in financing the project because the Supervisors did not sell enough of the bonds nor make funds available as fast as required by the engineer. Years later O'Shaughnessy talked about this at a 1925 meeting of the American Society of Civil Engineers (S.F. branch): "The result was that contractors often had to solicit banks to buy the bonds, as the City could not award a contract until adequate funds were deposited in the city treasury. This happened in the case of the railroad contract, in the subsequent contract for the dam at Hetch Hetchy, and in the tunnel aqueduct contract, thereby providing fuel for the political cauldron."

As men from his office went into the mountains on foot and horseback for final surveys, the City Engineer and his assistants made a number of important changes in the Freeman Plan. These changes added to the capacity of the project, worked for ease of supply and construction, and lessened expense to the taxpayers. Since immediate delivery of Hetch Hetchy water was not necessary and electric power was in demand, it was to the City's advantage to begin as soon as

The Groveland area could claim a population of about 300, including the surrounding ranches, when the above photo was taken in 1916. That's the schoolhouse in foreground, the Big Oak Flat Road proceeding down into town. The scene hadn't changed much when a similar view was taken fifty years later; there were now school buses and the highway had been paved.

Muddy Big Oak Flat Road passes in front of Groveland Hotel in this 1914 view at the east end of town. *(—Ernie Beck)*

San Francisco's surveyors going into the Hetch Hetchy region chartered stagecoaches from Charlie Baird in Groveland. Other parties hired stages to take them into Yosemite, since there was only infrequent service beyond Groveland. This party is leaving Crockers Station, westbound, on the regular Yosemite stage. *(—Gaytes Photo, Bancroft Library)*

possible to generate the power resulting from water development, so that revenue from power sales could be used to pay interest and redemption charges on bonds. It was therefore decided to build the Hetch Hetchy dam at first to about three-quarters of its ultimate height and thus develop about 60% of eventual reservoir capacity. An aqueduct to carry water westward would be completed as far as Moccasin Creek and a powerhouse put in operation at that place as soon as possible. The remaining parts of the aqueduct were to follow in time to have mountain water ready for delivery when Spring Valley sources would no longer be adequate, in the early 1930s, but no sooner than necessary in order to minimize the financial burden on San Francisco.

There were only about 300 people living in and around Groveland, and perhaps 270 in near-by Big Oak Flat, when actual construction work started in 1914. Travel to the area had been changed and vastly improved in July of that year, however, when Charlie Baird replaced his stagecoaches with a 22-passenger autostage connecting with the trains at Chinese. Now, instead of having to leave town at four in the morning, westbound travelers didn't have to be aboard until 5:30.

San Francisco men and equipment helped with extensive road improvements in the district. Even the unpredictable Supervisors appropriated $2500 toward construction of a new road up Priest Hill, with San Francisco and Stockton merchant associations pitching in. That job took over 18 months, with most of the work done by volunteers from Big Oak Flat and Groveland. Tools were donated; mining companies provided dynamite; Baird's stages hauled the volunteers; engineer George C. Donaldson did the survey. The county finally came through and completed the job, building a new bridge across Moccasin Creek at the bottom of this "first piece of scientifically constructed roadway in the county."

At this time Tuolumne County purchased the "Oak Flat Toll Road" and began improvements, including elimination of the South Fork covered toll bridge. East of Groveland San Francisco personnel laid out a new section of road with a 6% grade to replace the old winding 20% section. Groveland's tiny newspaper reported on April 22, 1916: "Men are going up today to start work at the City's sawmill in the Hetch Hetchy. They will fix the roads on their way up." Crews of City employees worked on improving the area's water supply and initiated a system of sewers in Groveland. Resurveys were made of townsites here and in Big Oak, because the 60-year-old mining surveys were far from accurate, it being impossible in many cases to determine the correct location of property lines. There were some disagreements, property disputes and the like, but they were surprisingly few and seemed to be settled amicably in that initial flurry of excitement. "Hetch Hetchy is coming" was the cry around Groveland and Big Oak Flat and in every way it was a loud cry of welcome.

Hetch Hetchy Railroad construction was started in 1916 at milepost 26 on the Sierra Railway. The first fill is shown above as the new railroad diverges from the Sierra (foreground). New crossties are stacked awaiting arrival of the first rails.

In March 1917, a year after start of construction, the contractor's storage yard at Hetch Hetchy Junction was crammed with stocks of ties and track supplies. Rolandi's engine No. 7, 4-6-0 type, was originally the same number on McCloud River Railroad and also worked for Weed Lumber Company in Northern California. *(—Both, City of S.F.)*

5

BUILDING HETCH HETCHY RAILROAD

Some means of rapid and comparatively low-cost transportation was the first and most apparent necessity to proceed seriously with construction work of this magnitude. There were two alternatives: a road adapted to heavy motor trucks or tractors, or a railroad. Analyzing costs alone, it was determined that a rail line would accomplish this at less cost per ton-mile, including the cost of its construction. Besides, such a system would earn something from outside passengers and freight – millions of feet of lumber were awaiting a carrier – and would remain in the end a very considerable asset. The comparison was purely academic anyway, because trucks wouldn't be able to operate at all when winter snows blocked the mountain roads, a condition that sometimes lasted until the end of April.

In 1912 the City prepared a map showing "Present and Proposed Railroads" for transportation to the damsites. First was the Hetch Hetchy & Yosemite Valley Railroad, a narrow-gauge logging operation of the West Side Lumber Co. This took off from the terminus of the Sierra Railway at Tuolumne City and snaked its way eastward 30 miles to Hull's Meadows. An additional 35 miles had been surveyed as far as Lake Eleanor. South of this and generally following the Tuolumne River, the map showed surveys for three standard-gauge "Hetch Hetchy" rail routes. W. Ham Hall had proposed a tough line going right up the canyon, while Marsden Manson showed a line following closely the old Yosemite Short Line survey, passing two miles north of Groveland, then down to the river near Lumsden Bridge and up Cherry and Eleanor creeks to the eventual site of Lake Eleanor Dam. The third HHRR proposal had been recommended by engineer Freeman. This one crossed the Tuolumne at Red Mountain Bar, about five miles below Jacksonville, went across the hills to Moccasin Creek, climbed Priest Hill and just touched the north edge of Groveland. Then it crept eastward a few miles along the very edge of the deep river canyon and gradually worked down until it ended at the bottom right where the dam would be placed at Hetch Hetchy.

O'Shaughnessy and his assistants accepted none of these, but took the best features of those following the river. They planned for the new railroad to branch off the Sierra line about 15 miles west of Jamestown, near Rosasco's ranch, go across country and down to the Tuolumne at Red Mountain Bar, then follow the river up to Jacksonville, go on to Moccasin Creek, then turn southward and begin to climb the face of Priest Hill. Groveland would be touched and from there the railroad would follow generally along the ridge eastward instead of making for river canyons. Thus the tracks would serve the thirty miles of main aqueduct east of the Sierra Railway, also Moccasin power development and a planned sawmill. Chief O'Shaughnessy figured that this higher route would enable his railroad to serve all parts of the project more easily and more economically. With only one exception, construction camps would be just a short downhill haul from the railroad, making a tremendous saving in time and labor.

The first actual construction was a nine-mile road from Hog Ranch to Hetch Hetchy, formerly accessible only by trail. Contract No. 1 was awarded to the Utah Construction Co. for the grading of this road late in 1914. Built suitable for a railroad roadbed, it would be 22 feet wide

Frederick S. Rolandi, contractor who built Hetch Hetchy Railroad, is shown in a San Francisco newspaper sketch with representations of some of his previous jobs drawn in above. *(—Frederick S. Rolandi, Jr.)*

so that it could be converted to highway use following construction years, as specified in the Raker Act. At about the same time an engineering party under Carl R. Rankin began laying out the rail route. Rankin had been an experienced railroad location engineer with the Southern Pacific and had surveyed the difficult San Diego & Arizona route. It took a year to lay out the Hetch Hetchy in full detail, Rankin being assisted by several engineers who thereafter remained with the project in other phases, among them Jack Best, Howard Tuttle, Frank Boothe and J. D. Cooper.

In 1915 there was serious talk of the railroad to Hetch Hetchy being built as a branch line by the Southern Pacific. O'Shaughnessy at that time said he would welcome such a possibility, while SP said they'd build if proper terms could be made. But apparently the big road couldn't foresee any good financial prospect for such a branch, particularly since the Sierra Railway would have to be taken over as well or serve as a bridge between SP tracks at Oakdale and the junction to Hetch Hetchy.

Rolandi used this Marion steam shovel at Six Bit Gulch to dig river gravel for ballast. On this occasion, photographed by fireman Waldo Bernhard, it had gotten off balance and tipped, being saved by the boom arm. *(—Waldo Bernhard)*

Piles of new rails and fasteners and the hundreds of other items needed to build a railroad fill the contractor's storage yard at Hetch Hetchy Junction a year after the work started here. The rail stocks look endless, but there were still 45 miles of track to be laid when this photograph was taken. Two years later the contractor was long gone and City crews had just finished this new station at Hetch Hetchy Junction, below. The Sierra Railway main line is to the left and Hetch Hetchy turning uphill to the right.

The initial crossing of Six Bit Gulch in the Tuolumne River Canyon was by this wooden trestle, shown during construction days. The contractor's leased engine, Sierra Ry. Shay No. 10, is shown with a work train. *(—Bud Meyer)*

Gravel for track ballast was obtained by the contractor from the bed of Tuolumne River at Six Bit Gulch. Spur tracks lead down to the Marion steam shovel. The main line heads up the river canyon in center distance. This entire area is now under the waters of Don Pedro Reservoir.

Track ballasting in progress in the canyon of Tuolumne River just prior to completion of the railroad in October 1917. Rolandi's inspection train with Shay locomotive No. 9 creeps slowly forward while workmen step aside. *(—Frederick S. Rolandi, Jr.)*

Melting snows high in the Sierra Nevada sent cascades of water down Tuolumne River Canyon in June 1916, while the Hetch Hetchy Railroad grade was being cut out of the cliff to right. Looking upstream toward Jacksonville from one-half mile below Woods Creek. *(—City of S.F.)*

Ten-Wheeler No. 7 of railroad builder Rolandi takes a train of supplies to construction camps in Tuolumne River Canyon. The covered car in center looks like a portable mess hall arrangement. Frederick S. Rolandi is standing beside the big light on back of tender. *(—City of S.F.)*

Railroad grading crews finally reach their first summit at the Divide, between Big Oak Flat and Groveland, in June 1916. The road is Big Oak Flat Road and the water at right flows down the last segment of famous Golden Rock Ditch, which here passed under highway and railroad in a siphon.

Another source of ballast for Hetch Hetchy Railroad was the gravel pit near Buck Meadows. The railroad's own crews obtained material here to finish ballasting after the contractor had completed his work. From the looks of things, some pretty large "boulders" found their way into HHRR right of way. The cars are lettered for "City & County of S.F. H.H."

After HHRR crews finished placing ballast the railroad presented a solid, well-built appearance in this mid-1918 view a quarter-mile below Big Oak Flat, looking down-grade westward. The track here nears the end of a long climb up Priest Hill and beyond. *(—Both, City of S.F.)*

A steam shovel digs out the railroad grade in Rattlesnake Gulch on the steep climb up Priest Hill. The heavy chain preceded steel cable in this type of equipment. *(—City of S.F.)*

Hetch Hetchy Junction was the name given to the actual starting place at milepost 26 on the Sierra. It was 40 air-line miles from there to the damsite, but it would be 68 miles by rail. On December 6, 1915, a contract was awarded to Frederick Rolandi, a former engineer with the Chicago, Milwaukee, St. Paul & Pacific Railroad and builder of several street railway lines in Shanghai and San Francisco. He agreed to build HHRR to Hog Ranch in 300 days for $1,574,000, the remaining grade to Hetch Hetchy Valley having been prepared by Utah Construction. About thirty miles of the entire route was to be through government lands (the National Forest) and the remainder purchased for about $30,000 from private owners, including ranchers, gold-mining companies and individual mine operators along the river. Ruling grade was to be 4%, compensated on curves, and curves to be not less than 90-foot radius — very steep and very sharply curved by standard railroad criteria. Steel rails were to be 60 pounds to the yard, laid on pine crossties. O'Shaughnessy noted that a Shay locomotive would be ideal on such an operation.

Construction started at the junction in February 1916, with W. H. Newell as engineer in charge for the contractor. The work proceeded eastward across rolling country for five miles before descending on sweeping curves into the Tuolumne River Canyon at elevation 600 (feet above sea level). The river was followed for eight miles to Moccasin Creek, and by April there were 900 men on the job. There was also a lot of trouble, local newspapers reporting on very bad labor dealings, violations of state sanitary laws on a wholesale scale in the camps, plus suits and threats from riverbank miners.

About a mile downstream from Jacksonville the railroad grade crossed the river by means of an imposing 220-foot-long steel bridge. Then it passed within sight of the town, but on the opposite bank, and now went about a mile along the old 1906 roadbed of the Yosemite Short Line. It crossed Moccasin Creek at elevation 680 and began the long climb up Priest Hill behind Moccasin. Up Grizzly Gulch it went, turned abruptly back on the opposite side, went left again around the shoulder of the hill, climbing endlessly until another eastward turn took it across Rattlesnake Creek and up this gulch to pass within sight of the famous old Priest Hotel. Straight ahead were Big Oak Flat and Groveland.

From Moccasin Creek a rise of 2390 feet was accomplished in 13 miles of continuous climb at a grade of 4% or its curved equivalent. Summertime temperatures along this portion of the route often reached 117 degrees, according to Waldo Bernhard, who was a fireman on construction trains. Rails expanded and warped so that there were frequent derailments on the hill before ballast was laid.

Passing right through an old Indian burial ground a short distance beyond Big Oak Flat, the line crossed over a divide at mile 26 (elevation

The railroad's "homemade" self-propelled tie tamper was put together with Ingersoll-Rand equipment under direction of Julian Harwood. He is shown to left in this scene near Buck Meadows; Harvey Femons, who made it work, is second from left.

Big Creek crossing, a few miles out of Groveland, shows unballasted track at time the railroad was turned over to the City by the contractor.

Fifteen miles east of Groveland, Hetch Hetchy Railroad skirted the canyon of Tuolumne River's South Fork. At this point, near Colfax Springs, the Big Oak Flat Road hung just above the railroad grade and the railroad hung nearly 2000 feet above the river. Work train with leased Sierra Ry. Shay No. 11. *(—Photo by Waldo Bernhard)*

South Fork station and camp at milepost 44. A mile down in the narrow canyon to left would be placed one of the project's major tunnel construction sites. Oakland Recreation Camp was just around the curve on the Middle Fork.

Jones Station, milepost 50, was an important spot for train-operating personnel, for here the camp cook was renowned and meals were excellent. The present highway to Mather runs on the old railroad grade from here.

Working eastward in August 1917, Rolandi's construction crews approached Intake station, 51 miles from Hetch Hetchy Junction. View from top of boxcar shows steel rails being dropped onto waiting ties; train moved forward as rails were spiked down. *(—City of S.F.)*

The last several miles of Hetch Hetchy Railroad construction took place in the heavily forested areas of eastern Tuolumne County. Here we see work proceeding at the Intake station area, from which point a cable incline tramway was dropped to Early Intake in the canyon. In the middle picture are new piers for the trestle over the South Fork of the Tuolumne.

In this historic photo by locomotive engineer Waldo Bernhard the last rails are being laid at Damsite, immediately above Hetch Hetchy Valley. This is October 1917, and construction foreman Leonardo Tomasso gazes proudly into the camera. *(—Waldo Bernhard)*

City of San Francisco officials, on an inspection trip, prepare to accept the newly completed Hetch Hetchy Railroad from contractor Rolandi (shown in light suit with arms folded). City Engineer Michael M. O'Shaughnessy is standing, wearing his dark suit and favorite black hat. *(–City of S.F.)*

3090) and was soon in Groveland, 27 miles from the junction. Townspeople greeted the construction trains with great delight. Tracklaying gangs came into town on a Sunday and every Poison Oaker in the area was out with his or her Brownie camera. The shriek of a shrill whistle on the work train's Shay locomotive was music to their ears, heralding great things for Groveland. That engine whistle had already gotten some of the town's schoolboys in trouble: Earl Pool recalled that five or six of them had cut school earlier in the week and run over to Divide to watch the train working upgrade from Big Oak. Eddie Phelan and Leonard Boitano were in the gang. It was a wonderful novelty to them, well worth the later punishment from their teacher. After all, hadn't the townspeople and ranchers met and encouraged City officials to build the railroad into town?

Rolandi used at least four locomotives to haul his supply and construction trains and personnel. Shays Nos. 10 and 11 were leased from the Sierra Ry.; enginemen Waldo Bernhard and Joe Cavagnero came with them. No. 7 was a 4-6-0 type bought secondhand from the McCloud River RR. No. 9 was a large Shay which the contractor acquired from the California-Western RR. Tom Connolly, who worked as a construction engineer for Rolandi, said there was also an engine No. 3, but no record of it has been found. The only other mechanical equipment used in building HHRR was two Marion steam shovels and a Rix compressor which drove one jackhammer.

The biggest problem in this operation was a supply of water for locomotive boilers, especially on Priest Hill. Engines often had to cut off and go back to Moccasin Creek, until track had reached Rattlesnake Gulch, where a water tank was installed. No. 7, the rod engine, did switching at Hetch Hetchy Junction and hauled supply trains out to Munn Siding at the bottom of Priest Hill. Beyond that point it was left to the flexible Shays, because No. 7's long wheelbase would have spread rails on the sharp mountain curves.

Supply trains ran night and day on account of the great distances involved and slow speed of the Shays. Dave Raggio, a Rolandi official, said a train would sometimes reach camp at 1:00 or 2:00 a.m. and then head right back. Engine crews carried their bedrolls along with them and often bedded down wherever the construction train happened to be; the only requirement was that they be near a supply of water and fuel. Jim Baker recalled that the fireman had to get up once or twice during the night to keep up a fire in the engine.

The railroad shop area would eventually be established at the east end of Groveland in a small meadow along Garrotte Creek. Railroad headquarters would be in the City's office building along

Face to face in the High Sierra, Hetch Hetchy's No. 1 meets Shay No. 7, leased during early operations on Hetch Hetchy Railroad. No. 7 was far from its home on the "Crookedest Railroad in the World." *(—Waldo Bernhard photo, Louis Stein, Jr., collection)*

Railroad builder Rolandi ran this final "mixed train" just prior to turning over the completed line to the City of San Francisco. His Shay locomotive No. 9, former California-Western No. 9, here pauses at the Six Bit Gulch water tank, while the passengers wait on their flatcar at the rear. *(—F. S. Rolandi, Jr.)*

Sierra Railway No. 10, leased for construction and early operation of Hetch Hetchy Railroad, was photographed by her engineer, Waldo Bernhard, with conductor Mose Baker sitting on the pilot beam. Bedrolls and luggage belong to the train crew; they were accustomed to bedding down for the night wherever they happened to be with the work train. *(—Louis Stein, Jr.)*

Four of the enginemen hired by contractor Rolandi during construction days on Hetch Hetchy Railroad. Mt. Tamalpais No. 7 was a familiar sight along the line in early days. *(—Waldo Bernhard)*

Below, the contractor's train of railroad-building supplies arrives at Groveland late in 1917, double-headed (or "double-ended"?) by Sierra Railway Shays Nos. 11 and 10. Railroad shops were later installed in the area in foreground and to left.

Hetch Hetchy Railroad right of way was blasted out of the cliffs almost 2000 feet above Tuolumne River. A few newly hired trainmen were known to walk right off the job after passing this area on their first trips! This view was taken shortly before rails were removed in 1948. *(—Al Rose Photo)*

the near-by highway. A passenger station was set up on the hillside just across the creek, approached by a footbridge which also served the hospital and a dormitory building.

East of Groveland the railroad followed generally the dividing ridge between the Tuolumne and Merced rivers. It first crossed Big Creek and the steel was laid on a shelf chopped out of shale alongside the stream, emerging from this canyon into an area of rolling pine forests. Hamilton Station was reached at milepost 39 (the name was changed to Buck Meadows in April 1916). Beyond here was almost all rugged mountain construction, the line swinging into the canyons of the South Fork and Middle Fork of the Tuolumne River and crossing each on a wooden trestle. Scenery was varied on a grand scale: glimpses of bare granite mountain peaks far to the east and frequent views of the deep green river sometimes almost 2000 feet below the track level!

Even before getting this far, railroad builder Rolandi was causing consternation among City engineers on the job. Much of the work was done very wastefully, according to Lou McAtee, one of the most capable construction engineers on the project. Carl Rankin sent letter after letter to Rolandi, advising him of material waste at various points. The contractor paid no attention, nor did he heed Rankin's suggestion that he order steel rails before the war affected the market. The price of steel went up and Rolandi was caught; at the same time he had to borrow and haul fill material from all over the place. But he went blithely on, grading, putting down the ties and spiking rails in place.

Near Jones Station, elevation 3810, the tracks reached another plateau. There was a steady light grade from here to Hog Ranch, but the new tracks twisted and turned as they followed the ridge on a ledge high above the main Tuolumne River. Early Intake was passed, but it was a camp out of sight at the bottom of that gorge to the left. Salt Log was next, then Ike Dye's place, and at last rails reached Hog Ranch at mile 59. From there, Rolandi's crews under Leonardo Tomasso rushed to finish the work over the prepared grade. At mile 62 the highest point was reached, Poopenaut Pass at elevation 5064. Six miles of continu-

Sierra Railway's Shay No. 11 was leased during the construction of Hetch Hetchy Railroad. Engineer Waldo Bernhard posed her here on the edge of Tuolumne River Canyon, then climbed down and snapped the picture. *(—Louis Stein, Jr., collection)*

ous descent on 4% grade completed the 68 miles to Damsite terminus, 3870 feet above sea level. The railroad was finished in October 1917; that is, the rails were all in place and trains could be operated — but not very fast and not too safely, for only a small portion of the gravel ballast had been laid down.

According to the Department of Engineering's annual report, since the contractor was unable to finish the railroad under the terms specified, a provisional acceptance was made, the City agreeing to complete ballasting and deduct actual cost from the final payment to Rolandi. But the latter surprised everyone by presenting a claim against the City, stating that all manner of unforeseen expense had arisen, that fill material was not available where promised and so on. The matter was eventually settled out of court, Rolandi picking up his final payment of about $195,000 in 1920. According to F. S. Rolandi, Jr., many years later, the contractor wouldn't accept a City voucher. He appeared with his attorney at City Hall and departed with four sacks of $20 gold pieces. They carried the sacks across the plaza, boarded the Montgomery Street "dinky" streetcar on Larkin Street and rode down to friend Giannini's Bank of Italy.

The total cost of the Hetch Hetchy Railroad came to two million dollars, but actual building cost was just a little more than $25,000 per mile, a remarkably low figure considering the terrain. And early use of the road enabled San Francisco to complete the dam at Hetch Hetchy in 3½ years, by April 1923. Thereby the City got priority to flood waters of the Tuolumne River.

Railroad officials stop their speeder to examine a kink in the track on Priest Hill. Such mishaps were frequent in the early days, caused by the effects of intense summer heat on the unballasted track.

Newly completed Hetch Hetchy Railroad in 1918. Above, the line crosses Rattlesnake Gulch on the 4% grade up Priest Hill. In a few years Priest Dam would be stretched across the background right behind this fill. Below, the railroad reaches Hetch Hetchy Valley in this view looking up the valley from above the dam construction site. *(–Both, City of S.F.)*

6

WE HAVE A RAILROAD!

If it had been possible the City and County of San Francisco would have avoided the chore of operating its own mountain railroad. Chief O'Shaughnessy not only tried to interest the Southern Pacific in taking the road on as a branch line, but approached the Santa Fe and the Sierra as well, also without success. So the City went into the common carrier railroad business, which it had to learn as it went along.

An early decision had eliminated possible operation by electricity, considered because of the war effort and fuel-conservation programs. Westinghouse and General Electric had both submitted proposals, the former recommending two 50-ton locomotives and the latter a single 70-tonner, although GE frankly stated that traffic would hardly justify the expense of electrification. Thus *Railway Age,* in December 1918, was able to say that the Hetch Hetchy was believed to be "the first steam railroad of any considerable extent to be built and also operated by a municipality."

The City engineering office appointed various of its personnel to administrative jobs on the Hetch Hetchy Railroad. O'Shaughnessy was named General Manager, with engineer Al Cleary as his assistant and later Superintendent. Subsequent Supers were to be the Construction Engineers in charge of Mountain Division aqueduct work: Nelson A. Eckart (whom the Chief called "my most faithful principal Assistant Engineer") and later Lloyd McAfee. The man who would actually run the railroad, as Assistant Superintendent and trainmaster, was Lester B. Cheminant. He was immediately dubbed "Chemmie" by the train crews.

For a railroad "of any considerable extent," the Hetch Hetchy was not very long, but it was a most difficult 68 miles for a standard-gauge common carrier. Of the total trackage, only five miles were level; going eastward, 42½ miles had a combined climb of 6333 feet and 20 miles made a total descent of 3415 feet. So there were long, steep grades in both directions. The line also had thirteen wooden trestles and two large steel bridges.

Steam locomotives require plentiful water supplies, and along the Hetch Hetchy there were eight redwood tanks of 100,000 gallons' capacity each. Since the engines would all burn oil for fuel, large steel oil tanks were placed at the junction and at Groveland, South Fork and Hog Ranch. To turn engines there was a wye at Hetch Hetchy Junction, a turntable in the shop area at Groveland and a loop at Damsite. Three additional wyes for locomotives would be installed as needed: one at Groveland to supplement the inadequate turntable, one at Moccasin and a former railcar wye at Hog Ranch. Gasoline-powered rigs would be prominent in the line's operations, and they were accommodated by small, compact wyes at Damsite and near the station for Early Intake. A light turntable for track buses and trucks was later added at the junction as tunnel work moved west; it was in front of a small "roundhouse" for the gas equipment.

When the Hetch Hetchy Railroad was accepted from the contractor in October 1917, the City rented two Shay locomotives, No. 7 was leased from the famous Mt. Tamalpais & Muir Woods Railroad, and No. 10 was the Sierra engine that had been used by Rolandi. Secondhand freight cars were bought during 1917-18 from a variety of sources and included nine boxcars, eight flats, four side-dump ballast cars and eight center-

Hetch Hetchy Railroad's track trucks were adapted from road vehicles to carry small shipments with great frequency. Above, Packard No. 9 with dump body. Right, Pierce-Arrow No. 8 carries workmen and their bedrolls to one of the camps. Note self-contained turntables mounted underneath 8 and 9. Below, Packard No. 6 with passenger seats stops at South Fork camp in 1918. The passengers are members of the Board of Supervisors on an inspection trip. Driver Jack Case stands on tie this side of truck. *(—Center, Bud Meyer)*

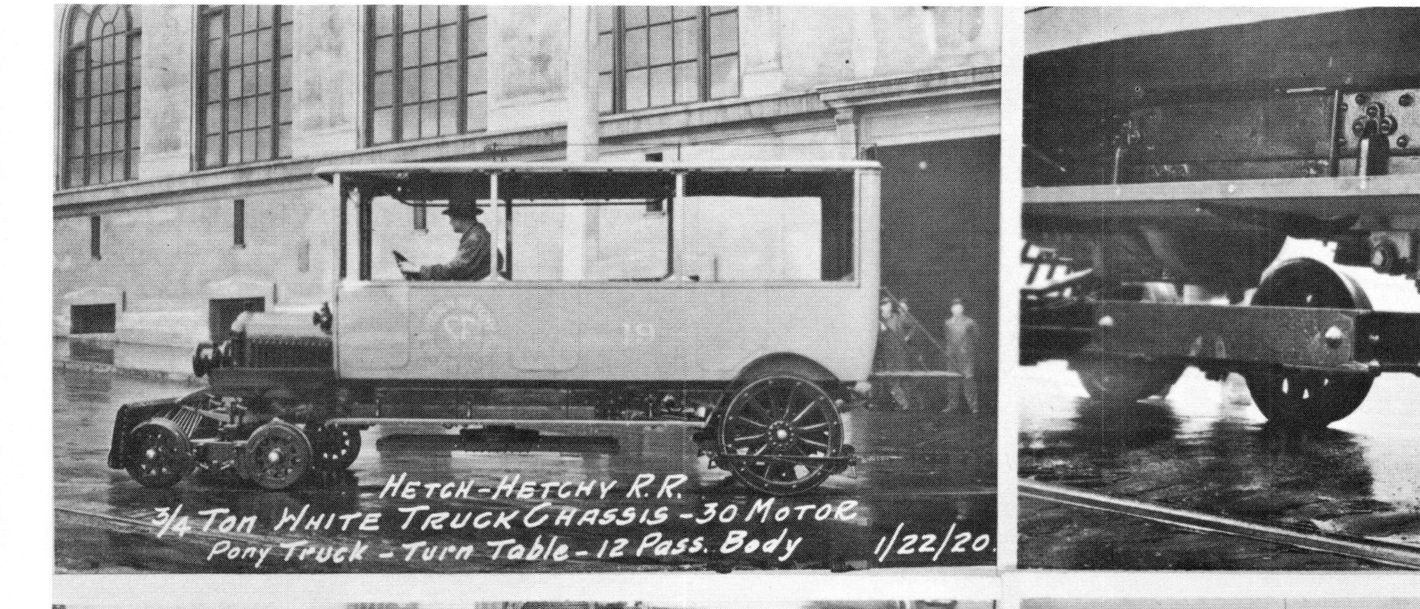

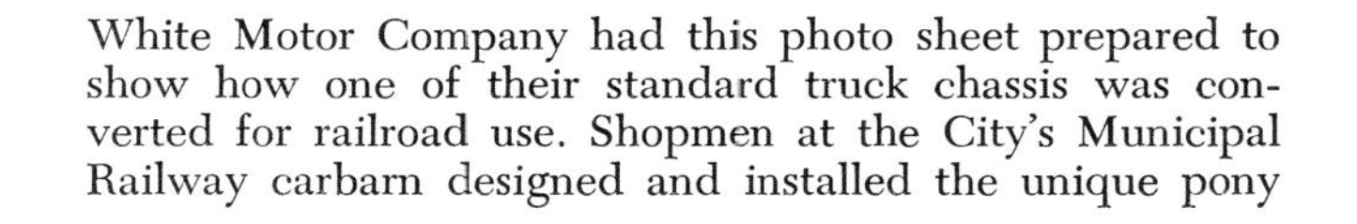

White Motor Company had this photo sheet prepared to show how one of their standard truck chassis was converted for railroad use. Shopmen at the City's Municipal Railway carbarn designed and installed the unique pony truck and braking system and mounted a turntable device underneath. The bodywork was by Meister & Sons, the car being used initially as an ambulance on rails. *(—White Motor Company)*

The first gasoline vehicle to be operated by the City on Hetch Hetchy Railroad was a converted Cadillac touring car, fitted out for rail operation in 1917 at the Municipal Railway's streetcar shops. It carried its own turntable slung underneath and the front wheels were originally left steerable so it could be removed from the tracks and driven short distances on the dirt roads in the mountains. Above, the vehicle poses outside the shops in San Francisco. The center photo shows this car in 1919 after a four-wheel pony truck had replaced its regular front wheels. This inspection party poses at Damsite, end of the line. From left to right: Mrs. Chaffee (wife of the official photographer), engineer R. P. McIntosh, right-of-way agent Joe Phillips, attorney Searles, State Inspector Gillespie, and Consulting Engineer John D. Galloway. Bottom view shows track truck No. 8, laden with everything from lettuce to brooms, ready to leave Groveland with camp supplies in 1920. Note that road license plate remains at rear beside the railroad lantern — the vehicle was readily convertible back to highway service. *(—Top and center, City of S.F.)*

Heisler locomotive No. 1 stops with a supply train at South Fork camp. To get materials from here down to the construction camp, a mile away in a deep canyon, there was a connecting "train" consisting of a small Cleveland tractor and a trailer on skids. *(–City of S.F.)*

dumps. A caboose was purchased and another later built at Groveland.

Operation as a common carrier didn't begin until July 1918, but the Hetch Hetchy was extremely busy before then. The first priority was to complete track ballasting, which entailed finding sources of suitable gravel. This wasn't always easy, and some fairly large "boulders" worked their way into the roadbed. Placing ballast was destined to be a continuing project for several years, as the management tried to complete the railroad and operate it at the same time. One innovation was a self-propelled tie-tamping machine put together at the Groveland shop in May 1919; with it over 500 feet of roadbed could be done each day. That job was finally completed in 1921.

Along with finishing the vanished contractor's work, the new railroad management had to speed up the hauling of an estimated 300,000 tons of construction materials necessary for building camps, dams and the aqueduct. Some of these projects were already under way. The Tamalpais engine began hauling camp supplies while railroad construction was still going on at the eastern end of the line.

Early in 1918 two more geared engines came to the Hetchy: Heislers No. 1, secondhand, and No. 2, fresh from the factory. They were a problem from the first, but there wasn't much choice, as flexible geared engines were about the only kind of power O'Shaughnessy could use on his unballasted track, especially in wet weather. Although original plans had called for three 2-8-2 Mikado-type locomotives, which would have been faster and easier to maintain, wartime scarcities and priorities made the asking price high ($47,500 each), and the ten-month waiting period was excessive. The new Heisler was only $24,100 with immediate delivery! So the railroad was operated exclusively with geared power until early in 1919.

Jim Baker started firing on the railroad while Rolandi was doing the construction. He stayed on after the City took over operation and was fireman on No. 2 when she arrived from Heisler. Baker recalled that they would start out with a train of materials and supplies from Groveland in the morning, run the 41 miles to Damsite and then get back as far as Jones for the night. At a Heisler's eight miles an hour this was a good day's work. These engines were designed for low-gear power rather than speed, but they had the bad habit, according to former railroader Ray White, of "dropping parts all along the line." Jack Best was more specific: they would "drop half their gears along the roadbed making a trip to Damsite and back." The management concluded that geared locomotives were fine on similar operations of not over 30 miles, but were too costly to maintain on the 68-mile Hetch Hetchy runs, where wear and tear "keeps locomotives in the shop about 50% of the time." And when they broke down on the line, what a headache! Red Fent

The original Groveland railroad "shop" was this siding alongside the main line. Heislers Nos. 1 and 2 are shown in this 1918 view, and the old Cadillac inspection car is having some work done at left. Beyond the water tank was the hospital, handy for emergencies but certainly not in a quiet zone.

maintained there was never a comfortable place to work on a breakdown: you were either freezing in the snow and ice or burning your hands on tools that had been lying in the broiling sun.

Snow was bad the winter of 1917-18, and the Hetch Hetchy wasn't prepared to meet that problem. Wedge plows were hastily improvised by Roadmaster Julian Harwood out of heavy corrugated metal and fastened to the engine fronts. But snow tended to pack against the banks with this sort of gear and then had to be shoveled away by laborers carried on the engines. When it got too bad, and before the snow could pack in behind them, the crews gave up and backed to Groveland. Sometimes they were tied up for days, playing cards and keeping the engine hot.

As early as 1917 the Hetch Hetchy Railroad put its first gasoline-powered railcar into service, and the story behind this piece of equipment provides an insight into how things were done in the road's construction days.

The Chief needed some sort of inspection vehicle to enable his engineers to keep an eye on construction and later to carry official parties from City Hall. The problem was turned over to Bill Bendel, in charge of the streetcar shops of the San Francisco Municipal Railway. Late in 1916 Bendel's men took a stock Cadillac and installed seats for seven and flanged steel wheels. The steering mechanism was left intact, as it was planned to use this car on both rail and road, changing from one to the other with a turntable built up under the chassis. By inserting jacks under both ends of the device, the vehicle could be moved off the track or reversed anywhere along the line in a couple of minutes. Bendel argued against leaving steerable front wheels, but he was overruled.

It was the responsibility of engineer Max Bartell to get this altered motorcar from city carbarn to mountain railroad. Many years later Bartell and Bendel described how they drove the auto down the Geary Street car tracks one midnight, derailed it at the appropriate crossing with a monkey wrench and headed down the rival United Railroad's car line on Fourth Street to the Southern Pacific depot. There it was shoved into a boxcar and transported to Hetch Hetchy Junction.

A four-wheel pony truck was eventually designed at the street railway shop and installed on the Cadillac, after the steerable front wheels had occasioned a number of derailments. The undercar turntable, designed and developed by Oscar Prinz of the City engineering office, proved so successful that duplicates were installed on three other track vehicles: two converted trucks and later railcar No. 19.

In early operations many supply and freight movements weren't large, and rather than wait for several days to accumulate enough tonnage for a locomotive and train, City engineers devised their own inexpensive freight carriers from gas-engine trucks fitted with railroad wheels, following the

No. 1 was a two-truck Heisler purchased secondhand from Dempsey Lumber Co. through a Portland dealer in 1918. Here she is shown at Damsite while Chief O'Shaughnessy and party prepare to depart by truck for the Hetch Hetchy Valley floor. *(–City of S.F.)*

During wartime the Heisler locomotive builder was the only manufacturer who offered immediate delivery, so Hetch Hetchy obtained a second engine of this type in 1918, three-truck No. 2. Power was transmitted to the wheels by means of two cylinders turning a central drive shaft; it made for high power but low speed. As all Hetch Hetchy locomotives were oil burners, the piled coal was possibly placed for this builder's photo. *(–Walter C. Casler)*

May 10, 1919: the permanent shop and yard area is functioning at Groveland; locomotives Nos. 1 and 3 stand outside the enginehouse-shop in early morning sun, ready for the day's work. A track truck, loaded with freight, waits alongside the warehouse to left. The main line, coming from Hetch Hetchy Junction, is the curved track to right. *(—City of S.F.)*

example of the successful Cadillac. One or two Packard trucks and a couple of Pierce-Arrows were equipped with standard railroad tires with 5½-inch treads shrunk over the original rims on the rear wheels; front pony trucks were built at the Municipal Railway shop. The Packards were 3½-ton rigs used mainly for carrying supplies, while the Pierces were a little smaller, one being fitted to carry 16 passengers. Ray Carne remembered Pierces Nos. 7 and 8 with dump bodies carrying ballast from the gravel pit at milepost 37.

Tiny Sheffield track speeders were used as "official cars" on the Hetch Hetchy from the earliest days. Construction engineers had them personally assigned for transportation between various work sites. They were sent out on emergency runs to take injured men to the hospital and often made special deliveries of urgently needed parts and supplies. At one time or another a "mail train" of one speeder would carry sealed pouches to camps beyond Groveland. Ultimately they were even used to carry passengers! When Doug Mirk went up from San Francisco to take a job as powerhouse operator at Early Intake in 1919, he stepped off the Sierra Railway train at the junction to be met by a railroad employee with a speeder, the "train" that afternoon. Upgrade out of the junction full throttle was turned on and it was scarcely touched again, uphill or down, until they reached Groveland, 27 miles into the mountains, only forty minutes later. Railroad boss Cheminant asked Mirk what time they had left Hetch Hetchy Junction. Mirk named the time and his "chauffeur" was fired on the spot!

Many former employees have said that Cheminant wasn't overly popular. He was something of an intellectual and a rather cold, distant man who took his responsibilities very seriously. The typical tough, professional railroad men didn't get along with him at all; on the other hand, those who got to know him found him honest, hard-working and a good friend. The opposition, of course, was delighted when he made errors, such as the time he gave himself clearance at the dispatcher's office to take a speeder west from Groveland. On a curve near the summit between Groveland and Big Oak he came face to face with the 2-spot dragging a freight train. There was barely time for Chemmie

Locomotive No. 3 arrives at Groveland, pushed up from the junction by Nos. 1 and 2, January 23, 1919. She still carries the number "1" of the Youngstown & Ohio River R.R. *(—Bud Meyer)*

Cadillac inspection car and Packard track truck No. 6 prepare to depart from Hetch Hetchy Junction with a load of San Francisco Supervisors on an inspection trip, October 1918. *(—City of S.F.)*

to stop and avoid killing himself. The boomers thought that was great and expressed regret that he'd been able to stop.

To connect Hetch Hetchy Railroad stations with the many construction camps and sites a variety of transportation devices were used: inclined railways at Damsite, Early Intake, Priest, Moccasin and Red Mountain Bar; an overhead cableway and narrow-gauge branch line to Brown Adit; rugged truck roads to Lake Eleanor, the South Fork tunnel camp and Adits 5-6 and 8-9. All the sites were downhill hauls from the tracks except the Second Garrotte Shaft, two miles from Groveland on the highway. Most of the road hauling was done by White trucks purchased in 1916 and 1918; they were similar to some of the later track vehicles and were famed for their endurance and longevity.

And so this most unusual and functional of railroads, employing several innovative practices and operated by City engineers instead of professional railroaders, was prepared as the "main artery" of the project and began hauling men and supplies so that the work might proceed.

Lumber for construction of camps, for forms, for railroad ties and trestle timbers was initially obtained from the City's sawmill at Canyon Ranch, near Hetch Hetchy. View shows logs being dragged along the skid road.

There was also a small sawmill at Lake Eleanor, making lumber for various construction requirements as the reservoir area was cleared. This October 1917 view shows mill in operation. Lake Eleanor was reached by road from Hetch Hetchy, the bridge to left carrying traffic over the Tuolumne River in Hetch Hetchy Valley. *(—Lower, Bud Meyer collection)*

7
PRELIMINARY CONSTRUCTION

A great deal of preliminary work, under way in some instances even before the start of the railroad, was needed to prepare for the construction of the dam and aqueduct. This work was accomplished under primitive conditions and in a working season kept relatively short by heavy winter snows and impassable roads. Once first snow fell, movement of equipment and supplies was almost stopped. Nevertheless, to get work started at Hetch Hetchy in 1915, the machinery moved in by road included a donkey engine and derricks, plus a complete sawmill.

Erection of the camp buildings at the Hetch Hetchy damsite was begun in September of that year on Hodeau Flat, an area that overlooked the valley from a point nearly above the site. A mile-long wagon road was dug and blasted from here down into Hetch Hetchy Valley on a 10% grade. Inclined cable tramways a fifth of a mile long were dropped 350 feet from the camp at railroad level to the river below. The main incline had a one-to-one drop on the lower portion and was powered by a 15-ton electric hoisting motor. In addition, wooden staircases were placed alongside the tramlines; some men would use them going down, but not often returning after work!

The job started with a 1000-foot diversion tunnel, inside the cliff at the south bank of the Tuolumne, that would carry the river past work areas at the damsite. Using hand drills on the stubborn granite, two crews put the 20-foot-diameter tunnel through and then went to work on a diversion dam which would turn the river into it. A smaller dam near the tunnel exit kept water from backing up into foundation excavations.

Alongside the rail route at Canyon Ranch, about five miles from Hetch Hetchy, was a fine stand of timber owned by San Francisco. The sawmill was placed here in April of 1915. Donkey engines with long cables dragged in the cut logs on skid roads. To preserve a natural forest appearance a "screen of trees" was left intact along the railroad grade. The cutting of lumber for permanent camps amounted to 1,200,000 feet that year. In charge was Ben Shaw, who had been boss of the big mill of the Yosemite Power Co. at Heardin's (Hardin Flat) in 1912 and was one-time superintendent of Golden Rock Ditch. A contractor also started the work of clearing the Hetch Hetchy Valley floor of all timber. Nothing was wasted, however, as trees were cut into cordwood and stacked, to serve later as fuel for donkey engines and dinky locomotives. By the end of June 1916, between 400 and 500 men were at work on the Hetch Hetchy Project.

To provide power for construction work a small hydroelectric plant was placed beside the Tuolumne River about 12 miles downstream, at the camp named Early Intake. Eventually water would first enter the aqueduct here and start its long, sheltered trip to San Francisco. But now, for water to turn the turbine wheels at this powerhouse, engineers went a few miles north to the Cherry River, where they installed a small diversion dam. Then, to ensure a year-round water supply so necessary for power generation, they went up and back to Lake Eleanor, reaching it by a hair-raising dirt road stretching 12 miles north from Hetch Hetchy itself. At Eleanor the first major dam of the system was erected – a beautiful, low, thin dam of multiple-arch construction, begun in Au-

To carry the water impounded at Lake Eleanor for power generating purposes, the small Lower Cherry Aqueduct was constructed from Cherry River to Early Intake. Initially, it consisted of tunnels, flumes and open ditch. Above, east portal of tunnel No. 3. Center, flume and wood-stave pipe where a break in the line had been caused by a landslide. The lower photo shows concreting of the open ditch; straddling the job and running on rails is a steam-driven concrete mixer. *(—Lower, Bud Meyer collection)*

A recent view of Lake Eleanor Reservoir and its dam, looking eastward toward the crest of the Sierra Nevada. The dam here was the first constructed on the Hetch Hetchy Project and the view to left shows the upstream face as this unusual multiple-arch structure neared completion in 1918. *(—Above, photo by Marshall Moxom, S.F. P.U.C.; below, City of S.F.)*

Above, carpenters at work in 1916 are building the box flume which would carry water to the powerhouse at Early Intake. A year-round supply of water for this purpose was supplied from Lake Eleanor Reservoir; its completed multiple-arch dam is shown to right in 1918. Below, pick-and-shovelers dig out the road into Early Intake Camp, February 1917. This finished road was so steep and narrow that most materials had to be lowered to the camp on an inclined cable tramway. *(—Above, City of S.F.; right, Gene Meyer collection from Bud Meyer)*

This small powerhouse at Early Intake provided electric power for all construction during the years when Hetch Hetchy Project was being built. Water to turn its generators came from Lake Eleanor by way of the Cherry River and Lower Cherry Aqueduct. Power was carried on wooden-pole lines to the various construction sites.

At right is the line gang in 1921; left to right: C. C. McMann, Shorty Mitchell, D. S. McCarthy, Frank White, Andy Townsend, Don Townsend. Below, looking down at Early Intake Camp in 1917; the powerhouse was placed a quarter-mile downstream, to the right, and water flowing to San Francisco would eventually enter the aqueduct tunnel a few hundred yards upstream, to the left. *(–Center, Don Townsend collection; bottom, City of S.F.)*

Powerhouse operators and their families at Early Intake, 1921. Left couple, Mr. and Mrs. Louis Hile; center, Doug and Muriel Mirk; right, Mrs. Max Mercer. Max was taking the picture with Doug's camera.

One of the City's 1918 White dump trucks on the grade near Early Intake. Dud Snider is in the middle and Everett Mallett to right. *(—Dud Snider)*

Project hospital under construction at Groveland, 1918. It was appropriately located next to Hetch Hetchy Railroad's main line, but it must have been somewhat noisy at times, as the tracks leaving town here were on a considerable grade for westbound trains. *(—City of S.F.)*

The City's sawmill at Canyon Ranch was alongside the railroad, five miles west of Hetch Hetchy. *(—City of S.F.)*

San Francisco's sawmill at Mather, 1921. When timber supplies were exhausted at Canyon Ranch, this larger mill was put in operation at Mather and provided all lumber needed on the project until completion of the Mountain Division work. The pond is now a swimming lake for Camp Mather. *(—City of S.F.)*

gust 1917 and put in service ten months later.

During those nearly impossible winter and spring months the terrible switchback road was conquered by trains of trucks, six at a time, carrying mostly cement and making three round trips each 24 hours. Four hundred of these "trains" did the job so well that the dam could be rushed to completion in time to hold much of the 1918 spring runoff. Thereafter, water released from Eleanor, combined with the Cherry's natural flow, was diverted by the small Cherry dam and carried three and one-half miles by Lower Cherry Aqueduct to the Early Intake powerhouse, turning three Pelton-Francis turbines. Each of these developed 1500 horsepower of electrical energy, which was then distributed at 22,000 volts to the construction sites on two wooden-pole transmission lines.

One power line extended 11 miles eastward to Hetch Hetchy, and the other 22 miles westward, following the line of the main aqueduct to Priest, with a two-mile branch into the Groveland headquarters. Serving 25 different points, these lines lighted camps and operated all construction machinery except steam locomotives and trucks on the road. This fine system was built under the supervision of Nels Eckart, assisted by electrical engineers Paul Ost and Andy Johns. Thus the location of this small powerhouse, Early Intake, was for a time the nucleus of the Hetch Hetchy Project.

The Intake station on the Hetch Hetchy Railroad was 16 miles west of Damsite, but it was almost at the top of the ridge, 1900 feet higher in elevation than the powerhouse and camp on the river. A very narrow and steep wagon road had been blasted down the mountainside in 1916, but it was entirely inadequate for heavy haulage of any sort. So here in 1918 was placed the longest inclined railway on the project: 3700 feet from the railroad down to the narrow canyon floor. It

The small "temporary" powerhouse at Early Intake has just been completed in 1918. Built to provide power for all construction work, the tiny plant actually continued in production until 1967.

The cliff-hanging road down to Early Intake on the Tuolumne River was so dangerous that this cable tramway was dropped on an incline from Hetch Hetchy Railroad, 1900 feet higher in elevation. These two 1918 views show the original arrangement with three rails, looking down (left) and up. One rail was later removed and the incline operated for most of its lifetime with a single car for freight and passenger service. *(—City of S.F.)*

The hoist house, with electric motor and winding drums for Early Intake's incline tramway, stood adjacent to Intake station on Hetch Hetchy Railroad. A one-inch steel cable 3800 feet long was required to lower the car to the camp below with its loads of freight and passengers. Maximum incline was 70 degrees, although this steepness is not apparent in the view below, taken from across the Tuolumne River in 1921. Between the buildings to left is the entrance to the aqueduct tunnel, from which excavated rock was brought to fill in the area at base of the incline. *(—City of S.F.)*

This cottage was provided at Canyon Ranch for sawmill superintendent Ben Shaw and his wife, shown on the porch in 1916. Shaw took over operation of the larger Mather sawmill when Canyon Ranch timber was exhausted. *(–City of S.F.)*

Wives and children of City officials at Hetch Hetchy were taken on a three-day "vacation" trip to Lake Eleanor in 1918, pausing for this picture on the way. *(–Mrs. Emilie McAfee)*

was 42-inch gauge and originally had three rails, the center one being common to both up and down cars. Later there was just a single track with one heavy car – heavy enough so that it could be used as a battering ram, dropped down on the end of its cable to clear snow in wintertime. A portion of the incline was on a 70% grade, and it's no wonder that a trip down this "devil's slide" afforded "one of the greatest kicks in this country."

By early in 1919 the sawmill at Canyon Ranch had turned out over 8,000,000 board feet of lumber. With this product they had built all construction camps, the Lower Cherry flume, a diversion dam, several buildings at Groveland, plus numerous railroad trestles, stations and even cars! The timber supply was then exhausted, and the sawmill was moved to Mather. (O'Shaughnessy had rechristened old Hog Ranch, naming it after Stephen T. Mather, the first Director of the National Park Service and later Assistant Secretary of the Interior. It was Steve Mather who had insisted that San Francisco leave in the mountains no scars that couldn't be eradicated.) The new millsite was beside a small lake in the forest, just a hundred yards or so from the Hetch Hetchy Railroad. Here again the skid-road system was used, donkey engines dragging in great logs of yellow pine, sugar pine, red fir and cedar. Redwood trees in the area were left untouched. Cutting averaged 25,000 to 30,000 feet per eight-hour shift, and by the time production was halted a few years later, 21,000,000 board feet of lumber had been turned out. So carefully was the work accomplished under Ben Shaw's supervision that visitors here today would have to search long and hard to find any evidence of logging and lumbering. That old mill pond is now a busy summer swimming lake for San Francisco's Camp Mather.

Hetch Hetchy Project headquarters was set up at the east end of Groveland. To the right, alongside Big Oak Flat Road, is the headquarters building. Across the way, to left is the Club House residence for single men, and to right the hospital. Railroad shops and yards filled the flat area in between and to the right. *(–City of S.F.)*

At the east end of Groveland, just beyond and across the road from the famous Groveland Hotel, was placed the City's two-story, gray and white headquarters building. On the ground floor were offices; part of the upstairs was living quarters for the Mountain Division Construction Engineer and his family. Steam heat was provided against the severe winters. Extending eastward from the building stood cottages for some of the married personnel. In a meadowlike area behind these structures were the railroad shop, freight house, warehouse, dispatcher's office and other buildings.

On the hillside beyond this railroad area, just across the main line, was a pair of sturdy two-story buildings: the hospital and the Club House, the latter a boardinghouse for unmarried employees. The 26-bed hospital was completely outfitted with an operating room, wards, private rooms and an X-ray room. Dr. Elisha T. Gould from Sonora was put in charge, with three "trained nurses" on duty at all times. To complete this San Francisco "colony" at Groveland were several new cottages higher up beyond the hospital. Dubbed "Silk Stocking Row," these were quarters for senior officials of the Hetch Hetchy Project. Water for the entire complex was obtained from old Golden Rock Ditch, repaired by City crews and leased from the Yosemite Power Co. for twenty dollars per month.

As soon as the World War was over at the end of 1918, men and equipment began flowing into the project again. The "south of the river" country stirred with the beginnings of boom times and natives were predicting a great future. Many permanent San Francisco employees were in residence; the long-awaited railroad was in operation; and there were plenty of jobs as dam construction and tunnel digging got under way.

The above scene is self-explanatory, but the setup, with flags and big bundles of bonds, indicates San Francisco's civic excitement on August 13, 1919. Just the day before, the contract had been signed for construction of the major dam at Hetch Hetchy, and officials from the City and from Utah Construction Company are shown (right) at the ceremony. Seated to left is M. M. O'Shaughnessy with W. H. Wattle of Utah Construction beside him; others in front row are Tim Reardon, D. G. Fraser, T. F. Boyle. Standing: J. B. Gartland, R. M. Searle, J. G. Tyler, A. J. Cleary, Ray Taylor, Ralph McLeran, W. J. Fitzgerald, Mayor Rolph, H. R. Smoot, J. W. Moyles, Dixwell Hewitt, Burl Armstrong, Joy Lichenstein. *(—Above, City of S.F.; below, Bud Meyer)*

8

THE BIG DAM AT HETCH HETCHY

On advice from the Army board of engineers in 1913, San Francisco found itself assuming leadership in providing water for all communities around San Francisco Bay – a total of about one million population at that time. Four million people would be living in the area shortly after the year 2000, it was predicted, and this would require the full 400 m.g.d. (million gallons of water per day) available from Tuolumne River sources. Fortunately, in the early twenties, the cities on the eastern side of the Bay pulled out of Hetch Hetchy and began developing their own supply from the Mokelumne. This change enabled the City over the years to meet constantly increasing requests for water from mushrooming suburban areas and industrial complexes stretching nearly fifty miles southward from City limits. Hetch Hetchy was planned and built with this in mind; additions can be made to various sections as needs arise. An increase in capacity can be made without changing the basic design.

The system was built as economically as possible, but where added expense was thought necessary to eliminate future outlays and endless maintenance, the work was done. The wisdom of this has been demonstrated again and again in recent years. But back in 1934 the cost of initial development reached about $100,000,000 by the time water flowed into San Francisco. Although the City had actually bought a bargain, there were many who did not agree and certain segments of the press had been on the attack from the very first day of the work. Journals published in the interests of the power industry predicted failure and empty pockets for the taxpayers. Some daily newspapers were obnoxious in their endless hammering, allegations of dishonesty by public servants, criticism of delays – anything to put the City's project in a bad light. Apparently the people of San Francisco were not deceived by these tactics, for over the years they voted several bond issues to pay for prosecution of the project. One of the major items they paid for was a big dam at the outlet of Hetch Hetchy Valley.

On August 1, 1919, a contract was awarded to the Utah Construction Co. for building this dam at a cost of slightly over $6,000,000. However, the City's 4½% bonds were not marketable at the time, and the Wattis brothers, founders of Utah Construction, had to arrange the sale of sufficient of these unattractive bonds before the City would award that contract. Then everything was ready; the railroad was operating; power was plentifully available; camps were built and supplied; and men were back from the Army and Navy. It was to take nearly four years of hard labor to finish this dam, called the "biggest structure of its kind in California." Some of the construction methods and details make fascinating reading a half-century later.

With almost perpendicular walls, the main part of Hetch Hetchy Valley was about a mile and a half long and two-thirds of a mile wide. The dam would block the Tuolumne River at a narrow gorge where the stream flowed from the valley. Water would actually back up a distance of seven miles, but beyond Kolana Rock the space was quite narrow. In prehistoric times this valley was gouged out and occupied by a huge glacier which carried large boulders along and deposited them in the "neck" where the dam would be placed.

In order to reach bedrock, engineers began excavating sand, gravel and boulders from the riverbed. Several large glacial potholes were found in the bedrock itself, some as deep as 15 feet. The lowest point for the foundation was at the bottom of one of these holes, 118 feet below the original streambed. Excavation was started with a steam shovel, which loaded narrow-gauge dump cars hauled by dinky steam locomotives. Material was deposited up the valley near a crusher site, to be used later in making concrete for the dam. As the shovel worked its way down, trains were forced to operate on an ever-steeper incline, reaching an "impossible" maximum grade of 20%. This was overcome by means of a cable counterbalance system whereby an empty train entering the pit helped a loaded one up the grade.

Below 65 feet, however, grades became too steep and space too narrow for steam shoveling. Now derricks were brought in, lifting excavated material out in skips and emptying these into trains of dump cars, the latter operating on trestlework along the sheer side walls, 40 feet above the streambed. The track on the south wall reached back into the valley by rails laid on top of the wooden diversion dam. Digging down month after month, Utah Construction would periodically say, "Well, it's deep enough." Chief O'Shaughnessy would hurry up from San Francisco, take a look, then say, "No. Go deeper."

Bedrock was finally reached several months after expected. Scale and soft pits were removed by sandblasting and polished rock surfaces were roughened. Loose sand and dust were removed with wire brushes and washing. Deep cuts had been made into both rock walls, from top to bottom, for anchoring the dam structure. The derricks placed large rock fragments and boulders on convenient corners and ledges for storage, until required as "plums" to be dropped into the cyclopean masonry. While all this was going on, engineers were building rock crushers and concrete mixing plants, elevators and sluices, and other machinery necessary for mixing and placing concrete. Sand and gravel were available locally; cement would come in over the Hetch Hetchy Railroad in 5000 boxcar loads.

R. P. McIntosh, who had come with O'Shaughnessy to the City job, designed the gravity-arch dam. He was assisted by R. J. Wood, with specifications prepared by Leslie W. Stocker, the Assistant Engineer who had done some of the original surveys back in 1908. McIntosh and Wood had also designed the multiple-arch dam at Lake Eleanor and were working on plans for later dams at Early Intake and at Priest.

The narrow-gauge trains which had hauled rock from the excavation were now made ready to carry aggregates back to be mixed with cement and thus form concrete for the main structure. This quite extensive rail system was the Valley Railroad; its "main line" extended four miles up the north side of Hetch Hetchy Valley to Rancheria sandpit, serving a rock quarry at the foot of Wapama Falls en route. There was one trestle and heavy grades which challenged the locomotives, for they were only small, wood-burning, four-wheel dinkies of the type used in construction everywhere before motor trucks were able to handle this sort of work. There were ten of them, mostly 18-ton Porters and a few Davenports, and to all but one had been locally added four-wheel "tender cars" to carry their cordwood fuel. The operators of these one-man engines were listed as "dinky skinners" and paid five dollars per day for running and firing their charges. Twelve flatcars and 92 four-yard dump cars made up the rest of the rolling stock of this tiny railroad.

Horace B. Chaffee, City Photographer, is the man historians must thank for the superb photographic record of building Hetch Hetchy, and particularly for his cliff-hanging views of this most interesting work at the damsite. It might be added here that commendation is also due to the City and County of San Francisco for seeing that all photographs of the project, and many of the negatives, were carefully filed and preserved.

Bulk cement was delivered in boxcars on a siding immediately above the job, unloaded into a hopper from which screw conveyors carried it to a storage bunker, and transported similarly from the bunker to two automatic weighing machines. These discharged by gravity through eight-inch pipes into two-yard Ransome mixers. Down in the valley gravel and sand were stored in huge piles under which were two railroad tunnels fitted with overhead mining dump doors on 14-foot centers. All 12 cars of a train could be loaded with mixed

Utah Construction Company operated the Valley Railroad, an extensive narrow-gauge system running from the dam construction site up into Hetch Hetchy Valley. Locomotives and other equipment for the line were lifted from Hetch Hetchy Railroad cars, swung out over the canyon on the Lidgerwood cableway and lowered to the narrow gauge 300 feet below. The tiny steamer here has had its water saddle tank removed to lighten the load. *(—Mr. & Mrs. Herb Whipp)*

aggregates without moving the train. Materials were then hauled to bunkers near the work and lifted by belt conveyors to charging bins above the mixers.

Placed just upstream from the dam face, the elevating tower was a tremendous timber structure 340 feet high and containing four hoists. Mixed concrete was distributed from here through various pipe assemblies. Spanning the canyon above everything was a 15-ton Lidgerwood cableway: 903 feet of 2¼-inch cable. It was situated to permit the placing of various valves as the dam was formed. All sorts of equipment — lumber, timbers, even Valley Railroad locomotives — were also handled by this cableway, which could drop materials from HHRR cars right onto the narrow-gauge cars 300 feet below.

Workmen and visitors were transported in the "basket," a wooden platform suspended from the cableway, as a form of rapid transit to working sites down below. A 1922 reporter described his ride as "the nearest approach to airplaning." Old-timers often told about the day timekeeper George Warren fell out. He dropped 125 feet and landed in a tiny, four-foot-deep pool of water at the base of the dam. They started sadly down to reclaim the body and found Warren only slightly injured! The Valley Railroad also carried sight-seeing passengers, just like any good tourist railroad today. Some groups of Yosemite campers, visiting the project, were taken right up Hetch Hetchy Valley on narrow-gauge flatcars fitted with seats. Visiting officials and various engineering groups were given this special tour as well.

Placing of concrete started in September 1921 and went on almost continuously until late spring 1923. Five hundred men worked at it in boiling hot summer and freezing winter and many lives were lost in this dangerous work. From December to March a heating plant was installed to prevent poured concrete from freezing. Often work went on around the clock and the nighttime scene must have been awesome in those early days of flood-lighting. Coming back from Lake Eleanor one reporter had the spectacular sight burst suddenly upon him as the road came over the northern mountain rim. It appeared a fairyland in the center of the Sierra blackness; no other word would fit. It seemed as though a million lights

Preliminary work at Hetch Hetchy Reservoir site in 1916. First a diversion tunnel had to be dug through solid granite in the south canyon wall; it would carry Tuolumne River water past the area while excavation and dam building were carried out. A Flory hoist is being lowered to the valley floor.

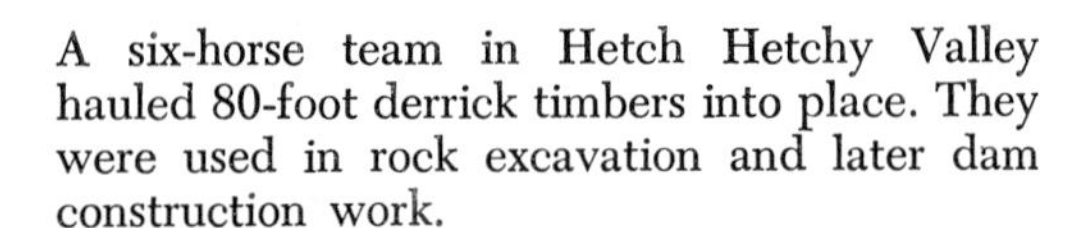

A six-horse team in Hetch Hetchy Valley hauled 80-foot derrick timbers into place. They were used in rock excavation and later dam construction work.

were driving back the night in that black gorge! "The crusher plant, the hoists and conveyors, the dinky trains – everything twinkles with lights. And above the dam, the great flood lights . . . as though it were broad daylight."

The last concrete was poured in April, bringing the total to 398,000 cubic yards. Height of the dam was 226½ feet above the original streambed and 344½ feet from the bottom of the foundation. It was 298 feet thick at the base and 15 feet at the top. A roadway ran along the crest, the length of two football fields, 605 feet. The dam was so constructed that 85 feet could be added to the height as necessity demanded, but it was already the second highest dam in the United States, Arrowrock in Idaho winning by ten feet. Behind the dam was one of the world's most beautiful reservoirs. In his 1933 report the California State Mineralogist made an interesting and intriguing comment: "The artificial production of a beautiful reservoir in Hetch Hetchy Valley may be taken as a compensation for the destruction by natural processes of the lake which in prehistoric times beautified the Yosemite."

O'Shaughnessy Dam was the name bestowed at the dam dedication, July 7, 1923, in honor of the Chief Engineer. Over 250 guests attended the ceremony, most of them coming by special train from San Francisco. Mayor "Sunny Jim" Rolph praised Chief O'Shaughnessy as "one of the great souls of our generation," dedicating his life and work to the community he loves. The City Engineer responded with gratitude, gave interesting details of the ongoing work and introduced his various assistants. The back of the Chief's hand was given to various critics who had said the work couldn't be done, who had thrown roadblocks in the way from the very beginning.

Flory hoist, derrick No. 1, was moved out of the riverbed in Hetch Hetchy Valley and prepared for use with a pile driver in construction of a diversion dam and other preliminary works of 1917.

There were two inclined cable tramways at Hetch Hetchy. They served as cliffside "elevators" on rails, transporting men and materials between Damsite Camp on Hodeau Flat and the construction work below, a difference in elevation of about 300 feet. The "New Tramway" is shown below in June 1920.

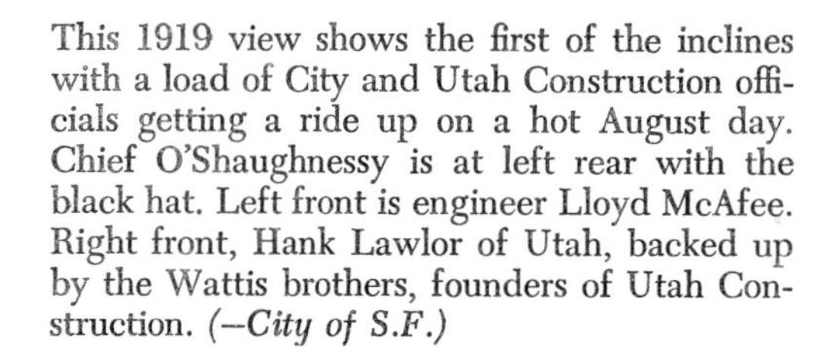

This 1919 view shows the first of the inclines with a load of City and Utah Construction officials getting a ride up on a hot August day. Chief O'Shaughnessy is at left rear with the black hat. Left front is engineer Lloyd McAfee. Right front, Hank Lawlor of Utah, backed up by the Wattis brothers, founders of Utah Construction. *(–City of S.F.)*

Preparatory work in 1916 was concerned largely with clearing the floor of Hetch Hetchy Valley for its use as a reservoir. Above, log boom across the river catches cordwood floated down to be stockpiled as locomotive and donkey-engine fuel. Shown to right are bunkhouses and mess hall of Hetch Hetchy Valley Camp for the crews working in the valley itself.

Four miles or so up the valley from the construction site was this bank of sand, needed in the concrete for the dam. Valley Railroad narrow-gauge train stands while the Marion shovel fills dump cars. The saddle-tank locomotives, normally carrying a small supply of coal or oil fuel, were here fitted with four-wheel tenders carrying wood for their fires. *(–City of S.F.)*

Rock excavated from the walls and floor of the gorge where the dam would be placed was hauled out to the floor of Hetch Hetchy Valley and stockpiled. Then, as needed in concrete work, it was loaded on Valley Railroad trains, taken to the crusher (in distance) and converted into gravel. The sand supply shown on page opposite was far up the valley, out of sight to right. *(—City of S.F.)*

Dinky loco No. 5200 (left) stands with a train of dump cars in bed of the Tuolumne River. As this area was cleared for dam construction, material was hauled out and stockpiled. The view below shows crusher plant, gravel and sand storage piles, from which measured mixes of aggregates were dispatched in trainloads to the damsite as required. Porter locomotives Nos. 3919 and 4010 are visible to far right; another is in the foreground near two Utah flatcars. *(—City of S.F.)*

At the dam construction site, south side of canyon, granite was cut and blasted out to make a trench for the dam abutment. The Valley Railroad's tiny trains were carried out onto both canyon walls by means of intricate trestlework. Utah Construction engines Nos. 344 and 5 are shown in the view above. Below, another view of the crusher plant in Hetch Hetchy Valley shows a stockpile of concrete blocks. As concrete was poured, these were placed within the dam itself to serve as drainage and inspection wells with openings 15 inches square.

North wall of canyon at construction site, looking upstream toward the valley. This gives a good idea of the extent of digging and blasting necessary to reach a solid footing of bedrock 118 feet below the original bed of Tuolumne River. In the distance are the two Hetch Hetchy falls, today visible from the dam in runoff season. *(–City of S.F.)*

Concrete mixing facilities at Hetch Hetchy were on the canyon's south wall. The large building at top left is the cement bunker, adjacent to Hetch Hetchy Railroad. From there bulk cement eventually came down eight-inch pipes to the mixers at bottom. Meanwhile, the aggregates (sand and gravel) were brought by rail and the proportioned loads dumped in left foreground. The inclined structure carries a conveyor belt on which aggregates were taken to bins above the mixers.

View looking upstream at concrete plant shows excavation for the dam foundation. The massive timber tower, to elevate concrete for placement, was claimed to be the largest timber structure ever erected, the top reaching 380 feet above water level (a building over 35 stories would be about as tall). At right is engine No. 344 with train of dump cars. In the distance, the bridge carries the road to Lake Eleanor over Tuolumne River. *(—City of S.F.)*

At bottom, deepest part of the foundation was 118 feet below the original streambed. Dam abutment trench runs up cliff face at south side of canyon. The tiny locomotive, with firewood tender and four dump cars, is operating on tracks at 3550 feet elevation.

At the concrete plant, pipes ran from the timber tower to carry concrete to various portions of the work. To right of tower is a staircase, but workmen were generally lowered and raised in the skip suspended on overhead cableway, which appears at upper right, hanging beside platform at Hetch Hetchy Railroad track. Below, carloads of large boulders were brought in as the dam was poured; a derrick lifted these big rocks and dropped them into the concrete throughout the structure. *(–City of S.F.)*

Upstream face of O'Shaughnessy Dam as the work neared completion, September 1922 above and March 1923 below. Locomotive No. 5 with train of aggregate cars crosses top of diversion dam. The river was carried past construction work through a tunnel at lower left. In the springtime view, below, the water was rising against the temporary dam. The road to Lake Eleanor, to right, was replaced by another crossing the top of the new dam when water rose in reservoir a few months later. Cars on Hetch Hetchy Railroad track appear at upper left. *(—Below, Chaffee Photo, City of S.F.)*

Concrete pouring goes on, six months before completion of the big dam at Hetch Hetchy. Concrete comes down chutes and pipes from the placement tower to be led to different areas. Large boulders and smaller granite rocks can be seen in place as "plums" in the structure, forming what was called "Cyclopean masonry." *(–City of S.F.)*

Looking downstream as concrete was poured in lower portions of the dam. The foundation was made sufficiently wide at this time to accommodate a later addition to O'Shaughnessy Dam without further excavation. *(–City of S.F.)*

maintained at one end of the rock pile. The capacity of rock storage is about 8,000 cu.yd. and the total sand storage is about 5,000 cu.yd. In loading the 12-car trains of 4-cu.yd. dump cars the train first backs under the sand pile and when the sand cars have been loaded the train is moved under the rock storage for the remainder of its load.

At the mixer plant these trains dump into receiving hoppers whence the material is elevated by belt conveyors to a 3-bin bunker. Each bin has a capacity of 250 cu.yd., the center one being used for sand and the two outer bins for rock. This plan makes possible the convenient arrangement of chutes direct from sand and gravel bins to each mixer.

The plant for handling cement in bulk which was described in *Engineering News-Record*, March 2, 1922, p. 352, continues to be very satisfactory. Each charge

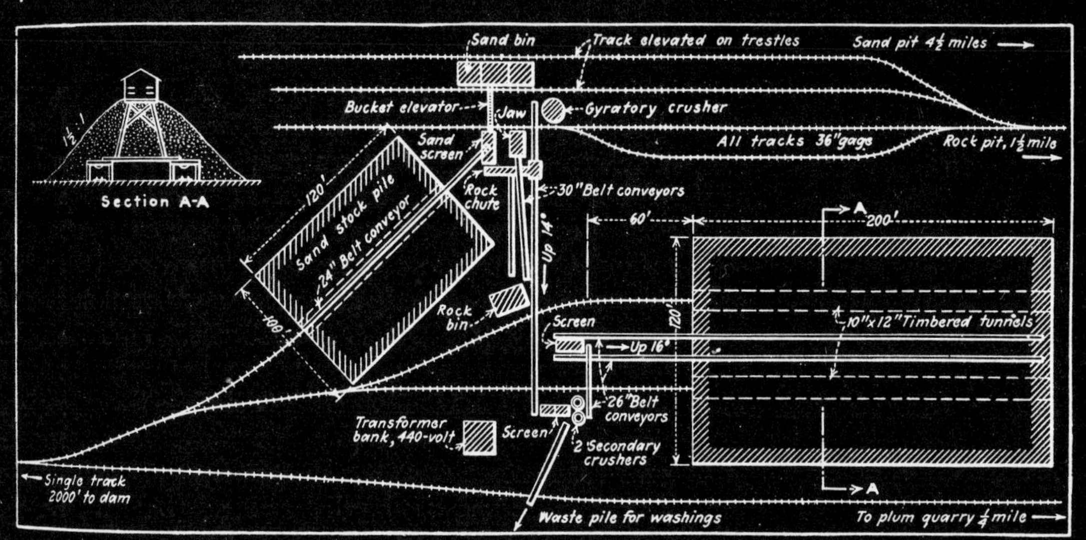

ARRANGEMENT OF EQUIPMENT IN MATERIALS PLANT AT HETCH HETCHY DAM

of cement is still weighed by hand as the automatic feature was not considered dependable. However, with an operator on each machine there has been no difficulty in delivering batches at an average interval of two minutes. With the proportions now used the richer mix takes 960 lb. of cement per batch and the leaner mix 799 lb. per batch.

There was some difficulty in getting this quantity of cement into the weighing compartment without delay and this was overcome by putting a ½-in. air pipe into the cement hopper just above the radial gate and directing its nozzle upward. By turning on the air when the hopper gate is open any tendency to arch or clog is prevented.

White, red or green signal lights are burned continuously on the weighing house to indicate to workmen on other parts of the job whether the rich or the lean mix is being delivered; red indicating for one scale, and green for the other. The same plan is followed at the head of the spouting system so the foreman on the dam knows when to change the point of delivery. A 1:3:6 mix is used in the body of the dam and a 1:2½:5 mixture is placed in the 5 ft. next to the upstream face, along the downstream face of the spillway and at certain other designated sections.

The cement from the measuring room is delivered to concrete mixers through two 8-in. steel pipes which drop the cement about 180 ft. on a 1 to 1 slope. During cold weather when hot water was used for mixing concrete steam rising in these cement pipes had a tendency to cause clogging. There was also the entrance of some rainwater at pipe joints. The joints were tarred to make them watertight and 8-in. vent pipes about 30 ft. long and curved downward at the upper ends were put into the receiving hoppers at the foot of the cement pipe to prevent the accumulation of pressure and to serve as a vent for water vapor. No pouring was done during zero weather. Cold weather limit was fixed at 20 deg. on a rising thermometer.

The mixing plant consists of two 2-cu.yd. batch mixers driven by 50-hp. motors and which ordinarily turn out 100 cu.yd. of concrete per hour. The maximum pour of 2,000 cu.yd. has been attained in a 16-hour run. The mixer operator manipulates all levers controlling the admission of materials into the mixer. With the exception of the water valve, all these gates are operated by water pressure. The sand and gravel measuring hoppers are located close to the outlet from the bin above so that when the measuring hopper is full it automatically chokes and the radial gate from the bin above can be closed at any time before the outlet from the measuring box is opened again.

Each batch is mixed a full minute. Charging and discharging occupy about another minute, so that normally a batch is delivered once every two minutes.

Each mixer discharges through a hopper that delivers to either of two skips which in turn deliver to a hopper which is changed to various levels in the tower at the head of the chute system as the work progresses. At this point a hopper capacity of about two batches has been found convenient in regulating the flow through the chute, in avoiding stoppages and maintaining continuous delivery. A separate chute system is used for each mixer, but the capacity of each chute is sufficient to take the output from both mixers whenever there is occasion for concentrating this delivery.

The surface of the dam on which concrete is being

A page from *Engineering News-Record* gave details of concrete preparation for the dam at Hetch Hetchy. The drawing shows narrow-gauge track arrangement at the rock crusher plant. For many years the Hetch Hetchy Project was a frequent subject in all sorts of engineering journals, and the City was fortunate in having several engineers who were better than adequate writers.

Making direct connection with Hetch Hetchy Railroad, the skip above traveled by overhead Lidgerwood cableway, carrying up to thirty men per load between the camp area and various working surfaces at the dam below. The carrying cable measured 2¼ inches in diameter. The average round trip took three minutes.

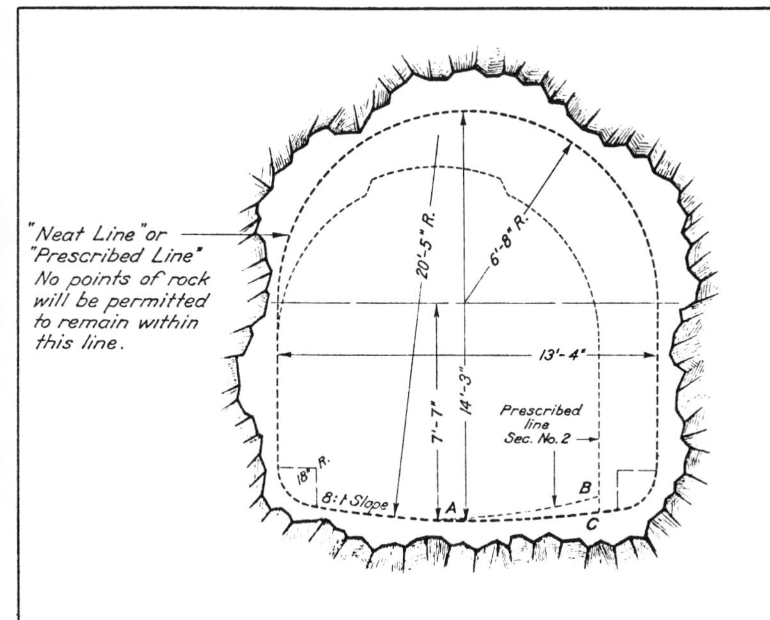

Water impounded behind the new dam was to be released in a year-round flow, first to generate electricity and then to flow on to the City's drinking faucets. Work on tunnels to carry this water part of the way was started and carried on during the same period that O'Shaughnessy Dam was under construction. Where tunnels in the Mountain and Foothill Divisions passed through solid granite, they were not lined. This drawing shows the cross section.

Dedication of the dam at Hetch Hetchy took place on July 7, 1923. At the ceremony it was given the name O'Shaughnessy Dam, in honor of the City's Chief Engineer, Michael M. O'Shaughnessy, whose energy, drive and careful planning brought the work to completion. The upper view shows part of the crowd, brought on a special train, gathered at Damsite Camp for the speeches. The center view shows Mr. O'Shaughnessy and Mayor Rolph standing by the plaque while Hetch Hetchy water gushes from the "perpetually flowing" drinking fountain. The lower view shows the "Chief" and the Mayor shaking hands, while various City and construction company officials look on. *(–City of S.F.)*

The completed dam, a few weeks before dedication. Note that in just a few months of spring thaw the reservoir has been filled to overflowing, excess water cascading down the stepped spillway on far side. *(—City of S.F.)*

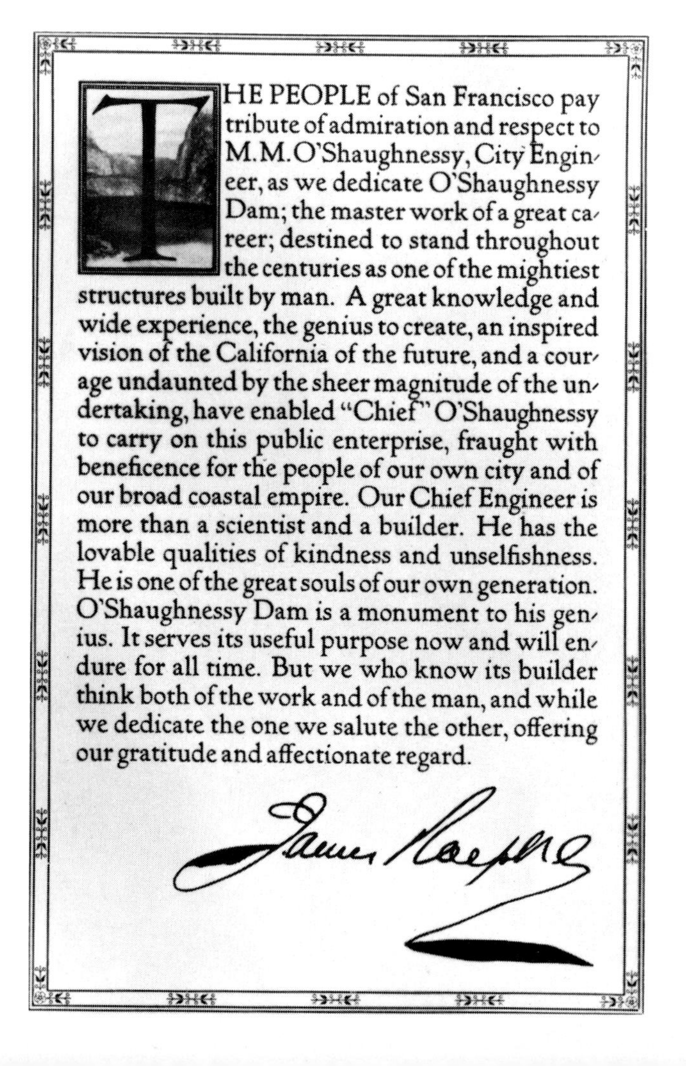

THE PEOPLE of San Francisco pay tribute of admiration and respect to M.M. O'Shaughnessy, City Engineer, as we dedicate O'Shaughnessy Dam; the master work of a great career; destined to stand throughout the centuries as one of the mightiest structures built by man. A great knowledge and wide experience, the genius to create, an inspired vision of the California of the future, and a courage undaunted by the sheer magnitude of the undertaking, have enabled "Chief" O'Shaughnessy to carry on this public enterprise, fraught with beneficence for the people of our own city and of our broad coastal empire. Our Chief Engineer is more than a scientist and a builder. He has the lovable qualities of kindness and unselfishness. He is one of the great souls of our own generation. O'Shaughnessy Dam is a monument to his genius. It serves its useful purpose now and will endure for all time. But we who know its builder think both of the work and of the man, and while we dedicate the one we salute the other, offering our gratitude and affectionate regard.

James Rolph

The words of Mayor James Rolph, Jr., were printed in the program for dam dedication ceremonies. It was Rolph who had pleaded with O'Shaughnessy to accept the job as City Engineer and Chief Engineer of the Hetch Hetchy Project. Ever popular, Rolph served as Mayor from 1912 to 1932, going on to be Governor of California. O'Shaughnessy served as City Engineer during the same period of time, then as a consulting engineer until his death in 1934. *(—Mrs. Emilie McAfee)*

Mountain Division Tunnel, first portion of the aqueduct, was started westward toward San Francisco in 1917. It was designed to carry 400 million gallons of water per day from Early Intake, below Hetch Hetchy, to Priest Reservoir above Moccasin Powerhouse. Much of the work was through solid granite, as shown in this view of the tunnel face at South Fork in 1918.

As rock was drilled and blasted out on a tunnel face, mucking machines moved up to load the muck into dump cars of the tunnel railway system. Drills and mucking machines were electrically powered; note the connections for wiring and the pipe carrying water to the drill bits. This 1919 view is at Priest Portal.

In order to get men, materials and supplies down to South Fork Camp, a good mile from the railroad down a very steep grade, this tractor "train" was utilized. The small Cletrac (Cleveland Tractor Co.) is shown climbing the grade with its sled trailer on steel runners. *(—City of S.F.)*

9

A MOUNTAIN TUNNEL GETS STARTED

From the new reservoir at Hetch Hetchy it was planned that conserved water would start the 149-mile journey to the City's storage reservoirs by flowing 11 miles down the Tuolumne as far as Early Intake. Here it would be taken into a long tunnel to begin its "enclosed" journey, the first 19 miles to another artificial lake at Priest. Designed to carry 400 m.g.d., the Mountain Division Tunnel was a mining job of tremendous proportions, 38% of it drilled through solid granite of such toughness that it required no concrete lining. Bids were invited in 1917 but were deemed excessively high on account of the war and were rejected. The City began the work itself, using "day labor." With hand-drilling methods, progress was very slow until the construction powerhouse at Early Intake was activated.

The start was made in summer 1917, with portals at Intake, South Fork and Priest. The first portion of the tunnel headed west from Early Intake, going four and one-half miles to the gorge of the South Fork of the Tuolumne River. Here in the canyon, hanging above the river in a maze of waterfalls and vertical drops, one portal aimed back eastward toward Intake, while directly opposite a second portal started westward toward Priest. This second part, over 13 miles long, was divided into five segments, so great care had to be taken that the various tunnel sections lined up both horizontally and vertically when crews met far underground. Two adits (side entrances) came into the tunnel line from the river canyon. Beyond that, where the line headed straight under mountain ridges, a pair of shafts were driven down from above. Thus, there were crews working at both ends of each tunnel, plus others driving the line in both directions from each of the four extra points provided by adits and shafts: twelve working surfaces in all when the job was in full swing, with men furiously drilling and blasting away at each.

All these hundreds of workmen had to be housed and fed; the City's construction camps were all built with the idea of comfortable housing and good food on the theory that men would stay longer where they were well cared for. Once things stabilized, very few people were living in tents. Most of the camps consisted of substantial wooden buildings, well lighted and ventilated and with stoves for heating. The relatively few married men were housed in separate cottages or provided their own quarters. After a few bad months at the beginning, Hetchy camps always were known for good cooks and plentiful food. Room and board was $1.25 per day, which was considered good. Several of the old-timers told the author that tunnel stiffs at that time preferred to work on Hetch Hetchy because of the good conditions. "The City had the best tunnel forces then working in the United States," said Owen Ellis who, as head of camp construction, was in and out of all the areas constantly.

On the Mountain Division Tunnel job there were construction camps at Early Intake, South Fork, Adit 5-6, Adit 8-9, Big Creek Shaft, Second Garrotte Shaft and Priest Portal. Intake was where that famous inclined tramway dropped down from the railroad into the canyon, near the small pioneer powerhouse. Before the tramway, Intake was a secluded camp reached only by a narrow, hazardous road laid out by Frank Booth where no road should ever have been built. The tramway

South Fork Tunnel Camp was precariously perched over stream and waterfalls at the confluence of the Middle and South Forks of Tuolumne River. Long building across the river is the cookhouse; engineering and timekeeping office is in front of it with the long staircase. Bunkhouses are on the cliff sides beyond in this view upstream from the tunnel site. Muck brought out of the tunnel headings was hauled around the canyon walls downstream and dumped, to be carried away in heavy spring floods. At left, battery locomotive shoves loaded dump cars out to be emptied. *(–Above, City of S.F.; left, Robert O. Richardson)*

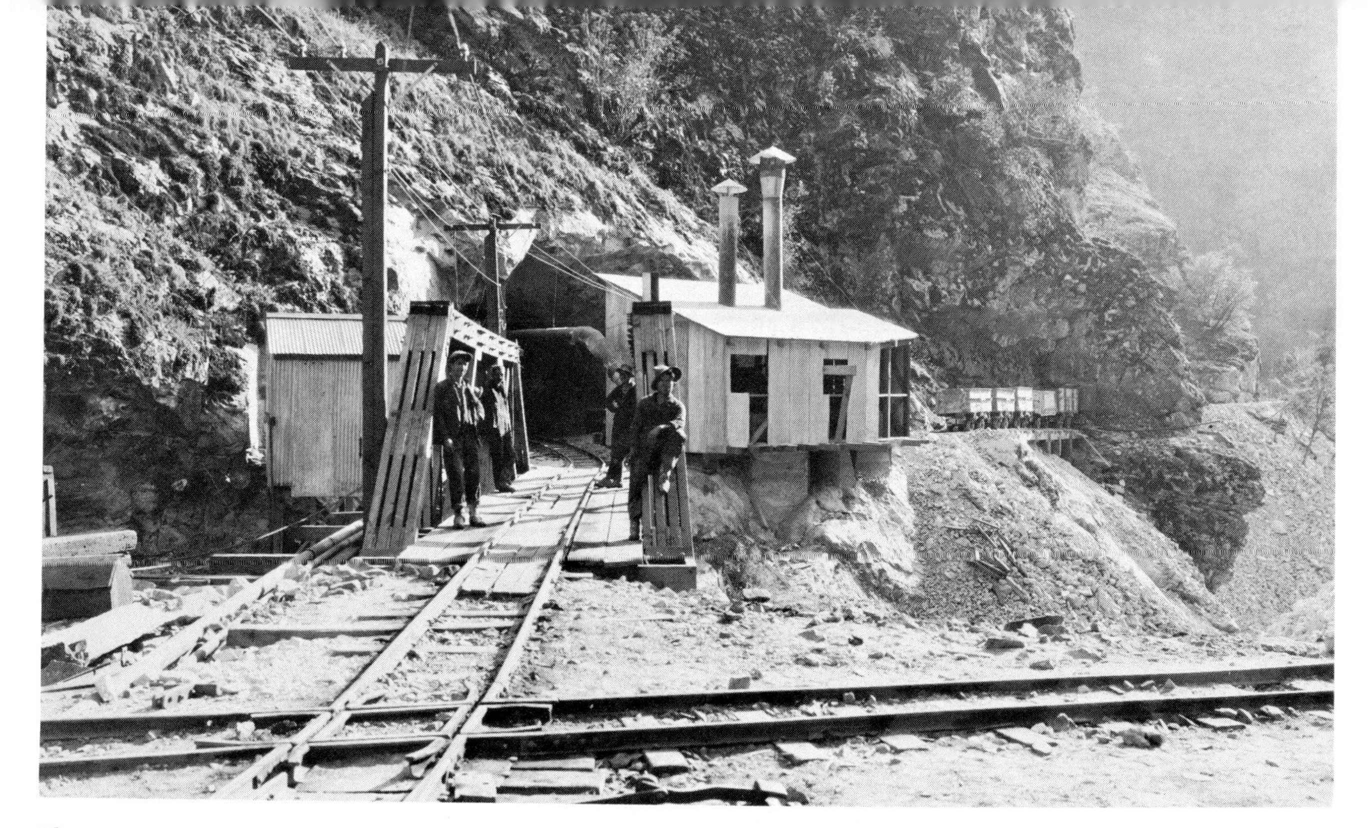

There were two tunnel headings at South Fork Camp, one going into the mountain on each side of the stream. This view shows the west portal, looking across the trestle, with dump area to right. Two-foot-gauge tracks of the tunnel railways fill the landscape. A hiker in the area today will have to search to find the slightest trace of any of these engineering works. *(—City of S.F.)*

was a blessing, the cable powered from above by an electric motor raising and lowering a heavily reinforced flatcar called a "strong back." All the workmen and their supplies and most of the tunneling equipment was let down to Intake on the incline: drums of oil, sections of steel pipe, shop machinery, food, later even bootleg booze. Inspectors came down, Supervisors and Grand Jurors from San Francisco, new men going on the job. People were hauled up to go to town on days off; kids were hauled up each morning for school; the sick and injured were hauled up en route to the hospital at Groveland.

Inside the mountain, miners under Tunnel Superintendent Matt Johnson were moving ahead into the granite 24 hours a day, making a hole 13 feet high and 13 feet wide in this unlined section. On the tunnel floor was a railroad track of narrow two-foot gauge, on which muck trains carried out tons of blasted rock and deposited it alongside the river. Powerful four-wheel, electric-battery locomotives did the hauling. About every hundred feet along the tunnel was an electric light, while big pipes were strung along the sides. One was a large blower conduit to supply fresh air at the cutting face and suck out foul air after a charge had been set off. The smaller pipe carried off water that drained into the tunnel from top and sides; since the tunnel was on a down grade, this water had to be pumped back and out the portal. A maze of wires and ropes was strung along with the pipes.

Riding into the tunnel in a wet and muddy muck train, a person immediately noticed the temperature change. In freezing winter months the air was much warmer than outside. But when the canyon outside was sweltering in summer, a visitor would come close to being chilled upon entering the portal. It was a bumpy ride in a semi-dark tube, providing a special kind of thrill to the observer. He held tight to the board seat, watching the cluster of lights far ahead get ever larger and brighter. At first, noise from the train would drown out that greater noise in the distance, but gradually as the scene of operations was approached a muffled thunder could be heard. The thunder grew to an almost deafening roar, awing most visitors.

Tunnel construction camp at Adit 5-6 was on the Tuolumne River Canyon wall a couple of miles north of Buck Meadows. Tunnel entrance is to right of dumping area. The long building houses machine shop, motor room, spare locomotive and perhaps mucking machine, locomotive battery charging area and supplies. To far left is the always-important cookhouse, with bunkhouses above. *(—City of S.F.)*

It was much the same at all twelve tunnel faces. The tremendous noise came from heavy air drills being driven into the virgin rock by muscular miners — mainly Swedes, for these men had a nation-wide reputation at that time of being the finest, sturdiest and most reliable tunnel men. A wet vapor surrounded the crews as they worked, so that they and their helpers toiled away in oilskins. Newspapermen, come to check and report for their editors and readers back in San Francisco, found this driving of tunnels the most fascinating feature of the project.

Most of the time the work was divided into three eight-hour shifts per day. In one shift the gang would drill a round of holes, then they'd be in the clear when the holes were loaded with dynamite and blasted. In 20 to 40 minutes the atmosphere would be clear, mucking (moving out the rubble) would start, and the next crew would be drilling away at the newly exposed face. Any shift that did more in its eight hours than the work allotted was paid extra, some earning bonuses as much as 40% above their wages. Ordinary pay, using 1922 as an example, was $5.50 per day for drillers, $5 for helpers and $4.25 for muckers.

To speed up removal of the broken rock, special machines were purchased. San Francisco's engineers had trained with air-driven steam shovels in their big streetcar tunnels, but these water tunnels were much smaller. Electric-driven Myers-Whalley coal mine mucking machines were tried out and proved highly successful. They were supplied to all camps except Big Creek and Second Garrotte, because at those locations machines had to be let down the shafts and smaller, air-operated Hoar muckers were used.

For all Hetch Hetchy tunnel work only the most rugged and least complicated types of mechanical equipment were selected. The primary consideration was reliability; next, simplicity and economy in operation, maintenance and repair; finally, cheapness in first cost. Each camp had a repair shop where spare machines and locomotives were kept ready to avoid delay. General Electric and Edison battery locomotives were supplied in four- and five-ton sizes, the battery being chosen over trolley for reasons of safety and convenience. Muck trains consisted of side-dump cars, some of steel having two-yard capacity and the rest of wood and steel with a capacity of three yards.

West of Early Intake about five miles was the next camp, South Fork, where tunneling was started in both directions from the river gorge. An upper camp was initially built alongside the Hetch Hetchy Railroad on the plateau above; then a rough, winding and very steep trail was dug out going down into the gorge a little more than one mile, a descent of over 800 feet. An incline wasn't possible here, so a supply train was made up using a Cleveland tractor (Cletrac) and

Perhaps the most important piece of tunneling equipment was the electrically operated Myers-Whalley mucking machine. Running on the tunnel railway tracks, the machine dug into newly blasted rock, lifting the muck up and over onto conveyor belts and so into dump cars at the rear. Photo at top shows one of these machines outside the shop building at Adit 5-6. Lower photo shows one of the smaller mucking machines, an Armstrong "Shoveloder," at Big Creek east heading. Where the machines had to be lowered down shafts, smaller equipment had to be used, the Hoar mucker being preferred. *(—Both, City of S.F.)*

At Big Creek Shaft (headframe to left) tunnel rock was brought to the surface and converted to gravel in this crusher plant alongside Hetch Hetchy Railroad. The material was used in track ballast and in county road improvement, with vast quantities stockpiled for subsequent use in concrete lining for the tunnel.

Below, the camp at Adit 8-9 perches on the mountainside. The view looks east, with the tunnel portal to right behind the trees. The river is to left. *(—Below, City of S.F.)*

Views at Big Creek Shaft, 575 feet above the tunnel line. Above, entrance to the three-compartment shaft in 1918, while the shaft was being excavated down to tunnel level. Center view shows the same area from opposite direction. Lower, the camp at Big Creek with railroad main line passing to left. Shaft was uphill about a quarter-mile to the right.

large box sled on skids. Three or four old fire department horses did "local" hauling around the lower camp, which was set up down in the gorge itself. Bunkhouses, mess hall, machine shop – the whole works – were crammed into the narrow defile, some of the structures actually straddling waterfalls and tumbling cascades, with canyon walls rising 600 feet above. This was South Fork Portal Camp which no sunshine reached for four months of the year, where in spring and early summer the snow-fed stream literally throws itself down between the confining walls. The first machine shop, built out across the river, was washed away in a few moments during a raging flood in 1921.

Muck from both portals was merely dumped into the canyon; there was no other place and the gorge drops steeply here. No matter how much rubble was accumulated during "dry months," the men were not particularly concerned about ultimate disposal. They had seen the force of this stream in runoff time and they counted on it to carry their muck on down the canyon. It did.

Pete Peterson was Tunnel Superintendent at this site. He became famous among Hetch Hetchy construction bosses, going on to push tunnel work westward on the Foothill Division and later the greatest tunnel of all, that through the Coast Range as the aqueduct neared San Francisco.

In the machine shop at each camp probably the busiest persons were the machinist drill sharpener and his helper. Drill steel was supplied in various lengths, ranging from 3 to 16 feet, with a hole running lengthwise through the center for water, which removed the cuttings and prevented dust. When cutting into granite, drilling began with a short length of rod with a bit 2½ inches in diameter and worked up to longer steel with a 1½-inch bit as the hole deepened. When a length became dull, it was sent back to the shop for sharpening. There in his forge the helper heated each drill end to about 1500 degrees and passed it over to the machinist, who could mold teeth to just the right degree of sharpness with the heavy, powered hammer of his Ingersoll-Rand drill sharpener.

Near the machine shop at South Fork, as at other camps, was the motor room. Here electric power was relayed from the power line to camp and tunnel; here also the blowers were operated. Usually, the battery for a tunnel locomotive would be on charge: these storage batteries powered trains for about six or seven hours each, the loco stopping in for a change much as a truck would stop at a service station for fuel.

Joining the two tunnels on opposite walls of the canyon at South Fork there would later be installed a huge 9½-foot steep pipe 225 feet long, on concrete piers. It would be by-passed by a U-shaped tunnel eventually, dug under the streambed nearly a half-century later.

Hetch Hetchy Aqueduct practically parallels the canyon wall of the South Fork and the main Tuolumne River for nearly five miles west from the crossing at old South Fork Portal Camp. The tunnel is just far enough in from the canyon wall to place it in solid rock. In 1917 seven adits were driven into the proposed tunnel line from gullies in this wall, in order to determine the nature of the rock, to give access to the work and to provide additional working surfaces. The adoption of mechanical mucking machines and electric locomotives led to the decision to carry on work from only two of these adits, those designated "5-6" and "8-9." The others served for ventilation and in some cases for dumping excavated rock. Camps, then, were laid out at 5-6 and 8-9 and mucking machines brought in to start working headings east and west at both locations.

So the next construction camp, two miles west of South Fork, was at Adit 5-6, near old Greek Camp. It was reached from Buck Meadows station on the railroad by going a mile and a quarter down a winding canyon road. The camp was a most spectacular place to live and work: way up on the canyon side, as on the side of a mountain, with the roaring wild river like a mild brook far below and grand mountain scenery "out in front." The place accommodated 125 men and was the pride of Shorty Calderwood, who had put it all together under the direction of Gene Meyer, "boss carpenter" of the whole project.

The tunnel line was about 120 feet inside the mountain from the 5-6 entrance. Gangs of hardrock miners were inching their tunnels eastward and westward under tunnel boss Andy Gallagher.

East heading of the aqueduct tunnel from the bottom of Big Creek Shaft. Timbering was necessary in areas of softer rock; these sections were concreted later. Pipe carries fresh air from surface to working face far ahead. Water was constantly seeping or gushing in from above – in some areas the men felt as if they were working in a rainstorm. *(–City of S.F.)*

One of the duties of the boss was to determine drilling procedures for each round of holes, based largely on the rock formation encountered. Sometimes 20 drilled holes would be sufficient, but other faces on the Hetchy job needed a full 42 holes. The first round would always be drilled toward the top of a fresh face, from a platform midway between invert and ceiling. Eventually each new face would have a set of 12-foot holes following the tunnel contour – the "cut holes." Then "secondary" holes and "lifters" were drilled in set patterns around and across the face, most of them straight into the rock, but certain ones set in diagonal paths toward a central point. When complete, all holes were filled with nitroglycerine powder to which electrical exploders were connected. Miners then left the area and the tunnel boss pressed a button to set off his explosion in a planned series of great booms.

Delay-action exploders account for the blasts being in a series rather than all at once. The first explosion might set off the "cut holes" or, more often, the diagonal "first relievers" at the center. Other explosions, perhaps a second apart, were designed to blast the rock in a pattern outward from center, then at the top of the face and finally at the bottom. As many as six explosions might be required in granite areas. Following these, vast quantities of rock would have been loosened and thrown back and poisonous fumes released. Blowers cleared the air; the mucking machine moved up with its flat, powered shovel; and broken rock was carried by conveyors to waiting muck cars behind.

Constant checking by engineers and survey teams was necessary to keep all those unconnected tunnel sections in line. To a visitor during construction years the tunnel would appear jagged and uneven, but under the superb guidance of "Sure Shot" Jack Best, location and alignment work was done to perfection. Best and Lou McAtee, Assistant Engineers, were lauded by responsible men years later as being the real builders of Hetch Hetchy.

From 5-6 to the next camp westward, 8-9, was a distance of nearly three miles. The only way to get there was to go back to Buck Meadows, take the road or a train to Smith Station, then follow another winding, cliff-hanging, three-mile road to the camp and adit. There 22 bunkhouses perched neatly on the hillside, with office, messhall and machine shop arranged near the portal. This camp was very much like 5-6, but the tunneling work under Superintendent Criddle presented a different problem entirely. The rock was of a looser formation and a lot of water was encountered; here the tunnel had to be timbered to hold until a concrete lining could be placed later.

The tunnel was widened at the bottom of shafts to provide repair and maintenance areas and some storage space while work was in progress. This scene is at bottom of Big Creek Shaft in 1921, looking east. Excavated muck was dumped into rock pocket below through grating to right; it slid underneath into skips, which carried the rock upward in the shaft, shown to rear of workman. *(—City of S.F.)*

At this location the aqueduct turns away from the canyon of the Tuolumne River and continues west for nine miles to Priest Portal at an average depth of 1000 feet below the surface. During construction it was reached by vertical shafts at Big Creek and Second Garrotte.

Big Creek Shaft was right alongside the railroad, just over two miles west from Adit 8-9. Hard-rock miners started this shaft in February 1918 and were able to begin tunneling 575 feet below 18 months later. It was a fairly large shaft with two hoisting compartments 4½ x 4½ feet and a manway measuring 4 x 4½ in which ladders and pipes were placed. The shaft extended 71 feet below tunnel level to a depth of 646 feet; this extra depth formed a rock pocket into which muck from both headings could be dumped. It would then be discharged into either hoisting compartment, skips of just over a yard capacity carrying the loads skyward. Excavated material was hauled at the surface by mule power to a near-by structure and crushed into sand and rock for concrete work up and down the line and for future use in lining the tunnel. In addition, a lot of the materials was supplied free to Tuolumne County to improve roads around Groveland. Many carloads provided ballast for HHRR roadbed.

Down in the earth at Big Creek the tunnel was widened to 16 feet where the shaft joined it. This extra width continued for 120 feet, to provide room for by-pass, locomotive repair and battery charging, storage and so on. The tunnel was driven in both directions by two eight-hour shifts per day, which were more economical than around-the-clock working. Additionally, the smaller, air-operated muckers here had to work the full 24 hours in order to dig out the material shot in two shifts of mining. Superintendent Fowler's men were dropped down and brought back after work in the same big buckets that lifted the muck. It was a dark, wet ride, for below 100 feet water forced itself through the sheathing like rain and no one went down the shaft without his long rubber coat and sou'wester. There was room in each bucket for about five men at a time, all holding on to the steel cable, falling water kept out by rubber wristbands.

A newspaperman, visiting here in 1922, noticed a difference: if anything went wrong, there would be no running or even climbing out! Everything depended on the winding machinery, cables and buckets. Everything was lowered in down the shaft and, until the tunnel was completed, that was the only way anything would get out: tools, mucking machines, locomotives and dump cars, timbering, dynamite, people, ventilation pipe. Hoar mucking machines were sent down in two sections as were locomotives and their batteries.

Second Garrotte area, showing shaft headframe to left, powder house to right. At this shaft a veritable underground river of water made work extremely difficult and progress exceedingly slow. Crews were at work 4½ years getting down 756 feet to the tunnel line. It is from this same shaft that growing Groveland today gets its fresh water. *(—City of S.F.)*

The second shaft necessary for working the big mountain tunnel was at Second Garrotte, almost three miles along the aqueduct line from Big Creek. This settlement was about two miles from Groveland along Big Oak Flat Road. Long years before Hetch Hetchy it claimed for itself fame as home of a couple of Bret Harte characters in "Tennessee's Partner" and had a well-photographed Hangman's Tree. Honest or otherwise, these and other attractions along the road to Yosemite had been publicized by an early-day stagecoach line. In this area the aqueduct went right through the famous Mother Lode gold belt for two miles. Mines were all over the place; in fact, there was a small gold strike at the De Martini place right back of the railroad shops in Groveland while the project was under construction.

Second Garrotte Shaft was the biggest headache by far on the Mountain Division portion of Hetch Hetchy. It was four and a half years before this shaft could be completed and tunnel digging started! The camp at top was a neat little layout, located in a meadowlike area surrounded by pine and oak trees. Everyone connected with this camp seemed to put a little something extra into his work, from the cook to Superintendent Bill Wiley and the Resident Engineer, John "Buddy" Ryan. Near the headframe was a pretty green and white cottage, built by City forces for the engineer, with all conveniences of a city apartment including lights and running water. The water was that fresh mountain stuff, right out of the shaft.

Water out of the shaft – that was the problem at Second Garrotte, and what a problem! Started in November 1918, the shaft encountered terrible water problems from the beginning. As it deepened, so the flow of water into it increased, going as high as 2000 gallons per minute at times, and a series of pump stations had to be installed in sumps at various levels adjacent to the man-made hole. It was as though an underground river had been tapped, for a sufficient quantity of water was pumped out to supply Groveland and all the ranches "downstream." As a matter of fact, for a long time the faithful pumps were boosting two million gallons a day to the surface so that work could go on. A steam pump was kept on stand-by in case of electrical failure. Lou McAtee said that during one year they were able to gain only sixty feet of depth while fighting off the heavy flow. He remembered that after one stoppage, a deep-sea diver had to be brought from San Francisco to drop down and clear the pumps! McAtee claimed that the whole basin was just a big lake covered with soil, water gathered over the years from seepage out of adjacent Golden Rock Ditch. With the ditch eventually closed, this saturation just disappeared (into the hole, up, out and down the creek).

Below 300 feet seepage was a minor problem, but of course that from above dropped down in bucketsful. Big Creek Shaft was dry in comparison. Yet the work was pursued relentlessly and with good reason. Chief O'Shaughnessy had esti-

Priest Portal, west end of Mountain Division Tunnel. Timekeepers, supply men, machinists and tunnel men in their oilskins pose by the sign proclaiming that they set the American hard-rock record, drilling, blasting and clearing 776 feet in August 1921.

mated that if as little as four months time could be saved by driving some of the tunnel east and west from this point, up to $600,000 could be saved. Each day sooner that the big planned powerhouse lower down at Moccasin could go into operation would mean $5000 more income from sale of the power to be generated. The tunnel level was 756 feet below the surface here and a further 64 feet of shaft served as a muck pocket. It was May 1923 when engineer Rawles was sent to Second Garrotte to start off tunnel headings.

When three shifts of miners were put to work at Garrotte, there were almost 250 men working out of the camp up above. For each heading underground, per shift, there were 17 miners, plus boss, walker, mucker operator and two motormen. Above ground there were three shifts of blacksmith and helper, hoistman, mechanic, roustabout and nipper to keep sharpened bits moving along. Miscellaneous workers included cooks and helpers, timekeepers and their helpers, and a few others. This was about standard for Mountain Division camps on the Hetch Hetchy Project.

The final camp connected with the Mountain Tunnel was at Priest Portal, western end and four miles from Second Garrotte. It was an exciting place to work: here the mechanical mucking machines were first tried out; here crews set a United States record for hard-rock tunnel drilling by advancing 776 feet during August 1921. At this same location along Rattlesnake Creek, other crews were building Priest Dam and clearing a reservoir which would receive the water and pass it on to the powerhouse in ordered quantities. Underground water flowing from Priest Portal was enough to supply the camps and to handle hydraulic sluicing for the upstream side of the dam; rock from the tunnel was carried out and dumped to form the downstream toe of this structure. The rest of the tunnel muck was crushed and stored for use in later concrete lining, scheduled to start as soon as all digging was complete for the entire 19 miles. Just above was the Hetch Hetchy Railroad, busy night and day with freight and passengers, its spur track joined to the tunnel camp by a new counterbalance tramway.

This rock train is emerging from Priest Portal. The motorman has his hand on the controller of a General Electric five-ton battery locomotive. There were two of these at each portal, one under charge while the other was operating. At the adits and shafts, where there were two headings, Edison battery locomotives were used, three at each location.

Underground water was not excessive in the tunnel at Priest and the rock was of such structure that blasting required only 25 holes per round. Unfortunately, there were a couple of bad accidents that spoiled the record; perhaps they resulted from pressure to get more footage and greater bonuses. Both were caused by miners drilling into "missed holes" – charges from the previous round that had failed to fire. The first accident happened at about 7:30 on the icy night of January 9, 1922. The Groveland doctor, John Degnan, couldn't be immediately located, so nurse Mary Meyer was rushed to the camp to administer morphine to survivors. Dead were Dick Sladden from the City and two Groveland men, Tom Ford and Frank Miller.

Four badly injured survivors were rushed to the system hospital, where Doc Degnan was now ready and started to work immediately. Mary Meyer remembered that he and the nurses worked until daylight before the final suture was tied. Some of the damage couldn't be repaired: W. S. McLeod lost both eyes, while Tex Lann lost his left arm and had a big piece of rock embedded in a hip. (Nurse Meyer said that a volunteer came in toward morning and asked if there was anything he could do to help. "Empty the garbage buckets," said Doc Degnan; the volunteer looked at the buckets, turned pale and walked out.) Lann was a huge man and the nurses were unable to lift him onto an operating table; so Tex insisted on doing it by himself. Also injured by flying rock were Julius Laetus and Enrique Tzars.

Old-timers remembered Tex Lann for many years afterward working for the City up around Moccasin. He got so used to having only one hand that he could even roll his own cigarettes with it.

On June 16, five months later, a second bad explosion brought death to a working crew. Angelo Segale and John "Mike" Haley died and the others were injured. And work went right on at Priest Portal and the other camps; accidents were expected, were part of the life of the miners. Some of the tunnel sections to the east were already being "holed through" as the Priest gangs drilled and blasted toward Garrotte.

For its day, the Hetch Hetchy Project was admirably sound from an ecological point of view, far less destructive of the environment than any comparable construction work done at that time or earlier. Trees were left standing to hide working areas; neatness prevailed at every site; rubbish was moved to hidden areas or buried. Much of the excavated rock was returned inside the mountains in the form of tunnel lining or employed in road work to cut dust and erosion. Such amounts as had to be left can only with difficulty be found today by the hiker or industrial archaeologist.

World Record broken by Construction Company of North America on Hetch Hetchy Tunnel, longest in the World

Lining Operations on 18-mile Hetch Hetchy Aqueduct Tunnel. City Engineer: **M. M. O'Shaughnessy.** Contractor: **Construction Company of North America,** San Francisco. Cement: **Old Mission Portland Cement.**

Pneumatic Concrete Plant at bottom of Moccasin Creek Surge Shaft, Hetch Hetchy Tunnel. City Engineer: **M. M. O'Shaughnessy.** Contractor: **Construction Company of North America,** San Francisco. Cement: **Old Mission Portland Cement.**

AT times its 18.3 miles (partly through granite, partly through compact slate) presented difficulties that strained engineering ingenuity. National and International Experts visited and studied the operations, simultaneously centered around 12 headings. In one section 700 ft. headway was made in 30 days!

Old Mission Portland Cement was used throughout the work. Mucking Machines and many other new methods helped to bring about this Engineering triumph; a credit to San Francisco, the profession, the Nation and the Construction Company of North America (Mills Building, San Francisco).

The exacting conditions of Hetch Hetchy Tunnel Construction called for a Super-standard Cement like

Old Mission

Portland Cement

"Sound as a Bell"

IT was used in all the Concrete work, notably the 18.3 miles of Tunnel Lining, preventing leakage or seepage despite gigantic water pressure.

Old Mission Portland Cement is made by the modern Wet Process and overtops Specifications by a maximum margin, in line with our policy to give the Building Public extra Quality without extra cost.

Old Mission Portland Cement Company

Standard Oil Building **San Francisco, Cal.**

Plant: San Juan, Cal.

Old Mission Portland Cement from San Juan Bautista played an important part in the Mountain Division Tunnel. For a time this was indeed the longest tunnel in the world and the tunnel men did set a record. But both of these records were soon surpassed in later stages of the Hetch Hetchy Project.

Dr. John P. Degnan was in charge of Hetch Hetchy medical services for many years, first at Groveland hospital, later at Livermore. He is holding the Hope baby, whom he had just nursed back to health. *(—Earl Hope)*

10
MOUNTAIN DIVISION TUNNEL COMPLETED

The Hetch Hetchy hospital, to which those injured at Priest and in scores of other accidents were taken, was one of the amenities of work on this project. Before San Francisco forces came into the area, sick and injured people had to be taken all the way to Sonora – 28 miles over mountain dirt roads by way of Priest Hill and Jacksonville. The City's hospital at Groveland, opened in December 1918, was a godsend, and San Francisco's compensation insurance rate was reduced through having this facility available. Doctors were on call for emergencies around the clock, special railroad ambulances speeding them to various camps in all kinds of weather. Every City employee on the job was charged one dollar per month for hospitalization insurance; people working for outside contractors were offered similar terms. Any person living in the area could be cared for at usual hospital rates. The service proved to be self-supporting. Dr. Elisha Gould had been hired to run the hospital at first but was soon succeeded by Dr. Homer Rose. Buddy Ryan remembered Rose as a fine man, "but he didn't like to go into the tunnels."

The legendary Dr. John Degnan took charge at the Groveland hospital in April 1921. Born in Yosemite Valley and known throughout the countryside, he was "Doc Degnan" to everyone. And he stayed with the Hetch Hetchy Project until the conclusion of initial construction in 1934. Engineer Lou McAtee many years later stated that Doc Degnan was the finest construction doctor that ever appeared on any job he knew of. Many remember the time Doc commandeered a "hot" locomotive at Groveland during a big snowstorm to rush off and deliver a baby at a lonely farmhouse far out of town.

One day in San Francisco Mary Isaacs, R.N., saw an ad in her newspaper asking for an industrial nurse. Having no idea what the job would be, she answered the ad and found herself at the Groveland hospital on April 1, 1921, the same day Doc Degnan was hired. Mary decided to work awhile, then get out of those mountains and return to the City. She never did. Instead, she met and married Gene Meyer, superintendent of camp construction, and she stayed on the Hetchy. It was Mary's opinion that Doc Degnan was something less than the finest medical man in California, but he had everyone convinced that he was a wonder and that was the main thing. All the workers on the project had confidence in Doc and his staff, and therein lay the success of Hetch Hetchy's medical facilities.

On July 27, 1922, the hospital at Groveland was destroyed by fire in the early morning. Thirty patients were moved to safety, although in getting the last one out nurse Ethel Moyer was badly burned, then suffered a broken back jumping from the burning building. The young nurse, 24, wife of a Hetch Hetchy fireman, died next day at Sierra Hospital in Sonora. A temporary hospital was immediately set up in the adjoining Club House, but it wasn't long before workmen had erected a new one-story hospital building with beds for 75 patients.

As we have seen, work on the 19-mile Mountain Division Tunnel had begun in 1917. Held back by limited funds, the job progressed slowly with day labor until February 1920. The market for securities was at a very low point and Hetch Hetchy bonds couldn't find a buyer. In fact, it was only O'Shaughnessy who kept the project alive in those very trying war and postwar years, ac-

After the Mountain Tunnel was holed through in November 1923, there remained a number of things to be completed before water could flow through the bore. A 9½-foot steel pipe, 225 feet long, had to be placed between the two tunnel portals on opposite sides of the canyon at South Fork. Sections were removed from railroad cars and winched more than a mile down to the site. *(—Don Townsend)*

A dam was required at Early Intake, to hold water and divert it into the tunnel. Work began on the arch dam late in 1923; when completed it was 81 feet high from the foundation and 262 feet in length.

Tunnel concreting machinery — a traveling setup of concrete gun and self-contained mixing plant assembly — rode on the tunnel rails as a unit about 150 feet long. Cars of dry mix were pulled up onto the elevated track, dropping their batches successively onto the belt conveyor, which carried them to the mixer at far end.

Rock crusher plant at Priest Portal. Tunnel entrance is to left out of the picture; cars came out and dumped rock on the trestle in foreground. The rock was pulled up the taller incline railway, dumping again at head of trestle-work, and so into crusher. Adjoining three-rail incline tramway was a counterbalance system connecting with Hetch Hetchy Railroad above.

cording to Ernie Rawles, who worked as a project engineer on various parts of the job.

Early in 1920 City officials wanted to see if they could now let out a contract for finishing this initial tunnel and have the successful bidder take care of financing sufficient bonds to cover cost of the work. Wartime conditions still prevailing, the City was advised to change its policy and receive bids on a cost-plus-percentage basis, with safeguards built in so that maximum stated costs would not be exceeded. When bids were opened in April, it was found that the total of guaranteed maximum costs under this type of contract were nearly 20% less than the total on a flat unit-price basis. So with a guaranteed cost of $7,802,952, the contract was awarded to the Construction Company of North America, which took over the City's forces and assumed direction of the work. The company agreed to take up City bonds in three annual installments, while voters approved increasing the interest rate from 4½% to 5½%.

The Construction Company of North America was organized just for the purpose of doing this job, with financial backing of the Old Mission Portland Cement Co. of San Francisco and San Juan Bautista. The President of CCNA was George Perry, and Vice-President, John A. McCarthy, who was also Vice-President and General Manager of Old Mission; C. Bruce Flick, CCNA Secretary-Treasurer, was likewise an officer of Old Mission. It is interesting to note also that the President of the cement company at the time was W. F. Humphrey, attorney and City Park Commissioner. Of course, Old Mission cement was used exclusively on the job.

Lloyd McAfee was the City's Construction Engineer on the tunnel and served as railroad Superintendent at the same time. He was assisted at Groveland by engineers Lou McAtee, Buddy Ryan, A. J. Wehner, L. B. Cheminant and electrical engineer Andy Johns. Engineer Al Cleary quit early in 1921, after 12 years with the City, to head up tunnel work for CCNA with Charlie Tinkler. When Tinkler went to Denver to bid on the Moffat Tunnel job, Buddy Ryan was loaned to Cleary to help finish the contract. CCNA apparently did little or no construction work other than the Mountain Division Tunnel; the company name had disappeared from the San Francisco city directory by 1928.

While brief litigation was pending on the CCNA contract, "Agitators from the East" invaded

As Mountain Division Tunnel was lined, collapsible forms advanced to the next section. Concrete was placed last on the bottom, or invert, as railway tracks were taken up.

the construction camps and instigated a general strike on August 23, 1920; it wasn't until the following May that full progress was resumed. There was a month's delay late in 1922 from another strike instigated by "radicals," according to the annual report. The City lost altogether nearly a year's time in the start of power generation. Chief O'Shaughnessy wrote Mayor Rolph that he was generally sympathetic with labor and the project got along well with the "regular unions" in San Francisco, but he wouldn't stand for outside unions coming in to intimidate the City and rob the taxpayers. It was O'Shaughnessy's opinion that Hetch Hetchy workers were among the highest paid in the United States at that time. In the earlier strike it had been the "Wobblies" (IWW); the later "outsiders" were the Mine, Mill & Smelter Workers. Whoever they were, they never proved a real threat. Buddy Ryan said that "Black Jack" Jerome could be on hand with the "usual lot of strikebreakers" from San Francisco at a moment's notice.

In 1922 progress was vividly demonstrated as tunnel headings coming from opposite directions began to meet and were holed through. The first was the tunnel west from South Fork to Adit 5-6 in July 1922. Then Adit 8-9 to Big Creek was completed on October 8, 1922. Six months later, on April 25, 1923, the tunnel section between Second Garrotte and Priest holed through and larger Myers-Whalley muckers were pushed ahead to replace Hoar shovels on the Second Garrotte east heading. In short order 5-6 was holed through to 8-9 on May 3, 1923, followed in two weeks by the long section between South Fork and Early Intake. Big Creek to Second Garrotte was last, the final rock crumbling and hands reaching through on November 26, 1923. Now crews could proceed with placing a concrete lining in the completed tunnel.

In San Francisco the building of lengthy and massive streetcar tunnels had enabled the City's engineers to study all factors that enter into concrete lining, including the economic factors. It was decided that about 11 miles of the new tunnel would have to be lined with concrete, leaving only the hard granite portions unlined. This work was done under a subcontract by W. F. Webb and Cox of the Universal Concrete Gun Co. Webb was the inventor of the gun, essentially a cylinder of ten-cubic-foot capacity in which a pneumatically operated piston forced mixed material through piping to the forms.

Collapsible tunnel forms were mounted on carriages (travelers) which moved ahead on the narrow-gauge track already in place. Also on the handy 24-inch tracks were the concrete gun and self-contained mixing plant assembly; feeding it were trains propelled by battery locomotives. During tunnel excavation, as rock had been brought out at Big Creek and Priest, it was crushed into gravel and sand, then stockpiled for the concrete lining. All lining materials were now taken into the tunnel at these two points alone, because they were directly on the railroad and eliminated use of trucks. Underground hauling for distances up to 7¼ miles was required, but it was accomplished smoothly by using two to five trains, depending on the distance, by careful location of passing tracks and by proper dispatching. Seven-car trains carried 21 batches of dry mix, three to a car, consisting of sand and gravel aggregates to which cement had been added.

As each train reached the mobile mixing plant far inside the tunnel, cars of dry mix were pulled up an incline onto an elevated track, under which a belt conveyor received materials from the cars successively, each of the batches being a load for the mixer. Water was added and the material mixed, then discharged into the attached 24-inch Webb gun. The concrete was shot into forms under an air pressure of 100 psi. Thus, concrete was placed within a few seconds after leaving the mixer, without waste and in prime condition. This wouldn't have been possible had it been mixed outside and transported into the tunnel.

The elevated track, mixer, gun and delivery pipe from the gun were connected together in a single unit about 150 feet long. It moved ahead as fast as the forms were filled, about 15 to 20 minutes being sufficient to move the entire plant the required distance. Ordinarily a form setup 120 to 140 feet long was made, about five hours being required for taking down, moving and setting up ahead. Then this amount of wall and arch would be poured in 12 to 16 hours. The system worked out very successfully. A record was set in August 1924 when in 29 working days a total of 4002 lineal feet of walls and arch were placed by one gun, requiring 7200 cubic yards of concrete.

After completion of the arch and sides the invert, or floor, was poured. In this way the original mining tracks could be used without change throughout the job. Now they were finally taken up, a section each night, as the night shift did preliminary work for each day's pour. The portable plant was moved back, track was removed, bot-

Priest Portal was cleaned up in 1925, after tunnel had been concreted and tracks removed. The view below shows pipe connecting both sections of the tunnel where the aqueduct line crosses South Fork. Recently this 225-foot steel pipe, considered a weak spot in the system, was replaced by an underground tunnel section. *(—Below, Oakland Recreation Dept.)*

A party of hikers from Oakland's vacation camp inspects the tunnel at South Fork in 1922. *(—Robert O. Richardson)*

tom cleaned and swept, and flowing water diverted past the next day's run.

Tunnel lining started on March 20, 1923; the last concrete was poured on May 12, 1925. Adits were closed with concrete plugs containing manholes for later entrance and inspections. The steel pipe joining both tunnel sections at South Fork was wrestled down into the canyon and put in place. Shafts were backfilled except for a three-foot-diameter concrete pipe opening. Finally the tunnel was complete; in the lined portion the finished diameter was 10 feet; in unlined granite, 13½ feet.

In order to divert Tuolumne River water into the aqueduct tunnel at Intake, it was necessary to build a dam — the fourth permanent dam of the system. Completed in 1925, it is a thin concrete arch 262 feet long and 81 feet high above bedrock. The water surface is automatically maintained at a constant elevation so that the required maximum of water is pushed through the tunnel to Priest. Designed for a flow capacity of 400 m.g.d., the tunnel can carry 470 m.g.d. under maximum flow conditions. An even greater flow can be attained if pressure is raised by increasing the height of water storage at Early Intake. The first water passed through from Early Intake to Priest on June 2, 1925.

Unforeseen events which occurred during tunnel construction had threatened to stop work on the project. During February and March 1923 the Intake powerhouse was shut down when a landslide carried away part of the aqueduct flume bringing water from the Cherry River. While repairs were rushed, power was supplied through a new stand-by tie with Pacific Gas & Electric near Moccasin. Then on December 8, 1924, while Mayor Rolph was away, work was stopped at several mountain camps and 500 men laid off on orders from the Acting Mayor, Ralph McLeran. The latter claimed that bond issue money was almost gone and he wanted to see O'Shaughnessy's authority diminished. A taxpayer's injunction held up the stop order, but two weeks were lost in finishing the tunnel. It would seem that neither side was without blame, but the Chief referred to McLeran's stop order as "cheap politics" in his book published ten years later.

A most unusual type of claim against the City and County of San Francisco was made even before completion of the tunnel; it spread and multiplied until in a few years there were scores of complaints. Landholders in the area between Big Creek and Priest maintained that springs on their property had dried up after the tunnel was dug through. They felt it had tapped their underground water flow. Jack White, who worked in the Hetchy office at Groveland, said one of his duties was to carry a barrel of water out of town every day to a rancher whose water had disappeared. Merle Rodgers carried water to various farms in a tank truck, according to Irene Magee.

But there hadn't been very much water in the district before the construction of Golden Rock Ditch and the farmers naturally suffered when it was gone; much of the area was again going arid. At any rate, the San Francisco engineers knew there was seepage into the tunnel and that water tables had been disturbed. All claims were fully investigated and were adjusted by 1927. On subsequent tunnel lines westward toward the City, engineers surveyed all surface water supplies before and after construction so that a firm basis for reviewing claims could be set.

Early Intake diversion dam and spillway, looking upstream. At extreme right is the house over the aqueduct headgates. The arch dam is 262 feet long. Until 1965 water for San Francisco flowed down the bed of Tuolumne River for 11 miles, from Hetch Hetchy to this point of diversion into the aqueduct tunnel. The new Canyon Power Tunnel carries this flow today, sending it through a new powerhouse just upstream from this dam. *(–City of S.F.)*

Hetch Hetchy Railroad headquarters at Groveland was east of the town, which is hidden by trees in center distance. The engine is No. 2; it's beside the shop-enginehouse, with its rear toward the car repair shop. Photo was taken from top of warehouse building, probably in 1919. *(—City of S.F.)*

The 1922 Groveland baseball team was made up largely of men on Hetch Hetchy payroll. Standing, left to right: Harry Rathbun, two unknown, Vince Lawrence, Lou McAtee, Bob Harder. In front: Walter Needy, Bob Scott and Joe Golub. *(—Photo by Lewis Hile, "Hetch Hetchy Photographer, Buckhorn Studio, Buck Meadows, Cal.")*

11
LIFE IN BOOMTOWN GROVELAND

It will be recalled that the area along Big Oak Flat Road and near Groveland in particular was pretty much dead to the world at the time work was started on the Hetch Hetchy Project. But from the opening of the "Hetch Hetchy Shaving Parlor," across the street from the Groveland Hotel, in February 1916 to about mid-1925 the place came alive with activity. Groveland found itself headquarters for San Francisco's construction forces and, incidentally, headquarters also for gambling, whoring, bootlegging and other excitements of a construction town in the Roaring Twenties.

The town had its sober element, of course. They lived their lives and ignored the rough element. Many Hetch Hetchy officials came from the City with wives and families; single men working in offices and shops lived at the Club House near the hospital. These people and the families of Groveland had their parties and dances and other entertainments, and in summer they all followed the baseball team. There seems to have been very little conflict between the rough and tumble amusements of Hetchy's miners in town on payday and the sober affairs of the more genteel. Rarely indeed was the sheriff called over from faraway Sonora. Women and children were safe; old people were treated with respect at all times. The growing kids thought nothing of seeing an occasional drunk asleep on the sidewalk of a Sunday morning. Those construction and railroad men worked hard and they played hard, but Groveland was able to survive it all quite well.

As the great project was built within and through the mountains of Tuolumne County, Groveland's post office became the second largest in the county by 1920. Forty miles eastward by rail the great dam was rising at Hetch Hetchy; the long aqueduct tunnel fostered teeming construction camps only a few miles out of town in either direction. Several local people were included in the workers at the project offices and railroad shops right in town, under an agreement the City made to employ a certain percentage of county residents in return for official co-operation. Groveland was now getting its water supply from the Second Garrotte Shaft; it was pumped almost two miles to storage tanks above town, mainly for service to Hetch Hetchy facilities. After this supply saved most of the town in a serious fire, residents formed a water users' association and bought the City's surplus. Similarly, San Francisco had put in a sewer line to serve its operations and invited the town to join in and help eliminate an unsavory situation.

It was only a very short time until Groveland boasted its own baseball team, made up largely of Hetch Hetchy employees and entered in the Mother Lode League. Among teams in the league (it varied from year to year) were those representing Sonora, Jamestown, Angels Camp, Melones, Stockton, Tuolumne City, Soulsbyville, Columbia and Oakdale. Chief O'Shaughnessy was a great baseball fan and loved to have teams representing any of his departments. He was especially proud of the Groveland team, started by his own personnel and starring mainly Hetchy workers.

Walt Magee, then a machinist in the railroad shops, claimed credit for getting the team organized in 1919, with help from Curley Gardiner, Willis O'Brien and Lou McAtee. Jack White of the office staff remembers that they fielded a

Scenes in Groveland in the early 1920s. The top photo looks east on Main Street (the Big Oak Flat Road to Yosemite). Large sign over the street in distance marks Hotel Charlotte, which is shown in the second view. The latter looks westward past Groveland Cash Store toward Hotel Baird, Groveland Garage and the stage depot. Picture below shows row of cottages along Main Street at east end of town; these were erected by the City and served as dwellings for married employees of the railroad, the shops and the office. *(—Top, Roy Brooks; center, Jim and Marie Neel; bottom, Moccasin Powerhouse collection)*

The City's office building, headquarters for Hetch Hetchy Project, was placed beside the highway just east of town; the view above also shows City barn in distance. To left, garage, stage depot and Charlie Baird's hotel; a name in the sidewalk is all that remains of the hotel. Below is "The New Hetch Hetchy Restaurant" with Groveland Hotel beyond. One of the signs reads, "Ice Cold Soft Drinks," a mark of Prohibition times (the picture was snapped in 1925). A corner of the Hetch Hetchy office building is in distance at far left. *(—Top, City of S.F.; center, Jim and Marie Neel; bottom, Roy Brooks)*

pretty good team from the first but had no diamond to practice or play on. A flat place on the Phelan Ranch just outside town was level enough but full of rocks and had been used to pasture the City's work horses. O'Brien and White got a Fresno scraper and a few helpers and "worked like dogs" to get the field smooth and in shape. They even built a small grandstand. A two-inch pipe was laid to bring water down from the railroad warehouse. McAtee, a good shortstop, was made captain and Bernice Laveroni, a local belle, served as team treasurer. For four or five years thereafter baseball season was the high point of Groveland's social calendar.

Practice was held two evenings a week, the team traveling down to their field on railroad speeders. They played a game each Sunday, even in the most intense summer heat. Often there was a Saturday night dance, which would continue till about five in the morning, with a midnight pause for food. They'd hurry home at dawn and try to get a little rest before the afternoon game. When temperatures got over 100 degrees, players would start to wilt about the third inning! A couple of times during the season games would be played with the loggers at Tuolumne City. Players, with their families and favorite rooters, would make the forty-mile drive after work Saturday afternoon. There would be a big dance, with a fried chicken feed at midnight, and eventually they would hit the hay. After the game on Sunday, the gang headed home by way of Sonora, stopping at the Inn for dinner.

Every year about the beginning of October the team would go up to Yosemite for "Indian Week," during which they played against various Indian teams – always good contests, because Groveland would then be playing without the "summer job" athletes from U.C., St. Mary's College and Fremont High in Oakland. These students, carefully selected for baseball ability, would be hired to work for San Francisco at Groveland during summer vacations. The work wasn't hard, but they were expected to play ball and hopefully lead Groveland to a fair number of victories. Some of the team were of professional caliber, having played in major and minor leagues or destined to advance far beyond the Mother Lode. Groveland had Hal Chase from the majors and well-known California ballplayers Pelegrini and the Cardosa brothers. Curley Gardiner, who came from St. Mary's to play in '22 and '23, went on to several minor league teams. Georgie Mathews later played for Oakland. It would seem that with all this talent the Grovelanders would easily wipe out the league and win a pennant.

Not so easy, said Bill Reed, for "the other teams had ringers, too," like Earl C. "Hap" Collard, who pitched for Stockton one year and for other teams from time to time. He went on to play for Cleveland and Philadelphia. Milt Steengrafe went from Sonora to the Chicago "Americans" and later Oakland in the PCL. At Sonora he got $100 if he won, $50 if he lost! The glamorous players were important, but the backbone of each team was made up of local men – miners, loggers, townspeople. All of the Groveland gang remembered that great ball team and its attendant excitement far better than they recalled details of the Hetch Hetchy Project.

No one could forget the tremendous rivalry between Groveland and Sonora. One year, 1923, Groveland won the first game 5-3 on their own field, even though Collard pitched for Sonora. There was a tremendous build-up of excitement before the rematch at Sonora. Groveland was empty, as a caravan of fifty cars had left on Sunday morning headed over the mountains to the county seat. With them they carried sirens, horns and lights rigged up with batteries, ready for the big victory celebration. Thousands of dollars were bet by Groveland supporters; Doc Degnan and a friend even walked up and down in front of Sonora fans, waving handfuls of greenbacks in search of yokels foolish enough to bet on the home team! Money was literally thrown onto the field after particularly good hits or spectacular plays.

Grovelanders never got to use their "special effects"; Sonora won easily, 9-3. What a shock! And then to be faced with that long drive back, up Priest Grade and all. Members of the team soothed themselves by stopping at Jacksonville for a fine Italian dinner and bottles of comforting "Dago red" before hitting the hill climb. They drank so much wine, in fact, that most of the team – that is, the Hetch Hetchy employees – came

Fourth of July in Groveland, circa 1922! Decorated autos began lining up early in front of the City's headquarters, ready for the big parade through town and back. *(—Mrs. Fred Cassaretto)*

Here comes the band! Five-year-old Floyd Cassaretto, riding his carefully groomed donkey, leads the parade on Fourth of July. In the afternoon there was a major ball game against another Mother Lode League team, probably Sonora, in the field down at Phelan's Ranch. *(—Both, Mrs. Fred Cassaretto)*

The Hetch Hetchy "office gang" at Groveland went for a railroad snow trip one crisp Sunday in February 1923. Left to right, top row: Ray Carne, Marie Floyd, Hilda Walsack, Jack Case, Joe Quast (leaning out window of bus No. 24); center row: Mrs. Quast, Mike Cowan, two unidentified, Grahame, Bob Hardy, Bob Cheminant, Trainmaster L. B. "Chemmie" Cheminant; kneeling in front: Fred Arndt and Mrs. Edith Cheminant. *(—Mrs. Edith Cheminant)*

very close to being fired next morning. The Sonora paper reported that "the boys from across the river spent 24 or 25 hundred dollars in Sonora during their visit."

It was a good rivalry while it lasted. Apparently 1923 was the last year for Groveland, as the big dam and other phases of the project were completed and crews moved westward. Toward the end of the season, after Groveland dropped out, a critical game was set between Sonora and Tuolumne. The latter team paid Al Kyte and Bill Reed to come over from Groveland and spent $100 to bring in a widely heralded pitcher from U.C. in Berkeley. Reed laughingly recalled that the great pitcher didn't last one inning and Bill was moved over to take his place at the end of the second, when Sonora was ahead 9-0. The game ended 13-9 with Tuolumne on top.

Truly, as Bill Reed told it, "they went mad over baseball" up there in the mountains during the years of Hetch Hetchy construction and afterward. It was the only scheduled outdoor amusement. But it was certainly not the only "fun thing" going on, as any ex-member of the famous Groveland Social Club will attest.

Jack White recalled that at night the single fellows, including himself, Harry Rathbun, Walt Needy and others, used to patronize various bootleg joints, their favorite being a tent known as "The Rag Dump." Hoping to provide better things to do and keep the office force out of the "dumps," accountant Willis O'Brien paid a visit to Silk Stocking Row up behind the hospital. There he talked the officials' wives into organizing what they called the "Groveland Social Club." At first their plans didn't interest the single men at all, but as soon as the ladies were able to get the girls to join, the inevitable happened. Nearly all Hetch Hetchy people joined and Susie White was chosen first president. Meetings at Odd Fellows Hall were planning sessions for card parties, trips and some of the greatest dances of the Roaring Twenties.

Among the organizers were Marie Floyd (she later married timekeeper Jim Neel) and Hilda Walsack from the City headquarters building. They got colored posters printed and spread all around town. Bands were brought in from Oakdale or Sonora, providing waltzes, polkas and fox trot music all night long in a hall across from the Groveland Hotel. Engineer Tom Connolly, later a well-known western construction man, said that he often walked or trotted from Hog Ranch to the dances and walked back to his station on Sunday, if there was no extra train or speeder he could grab a ride on. It was 32 miles each way! Many walked in from closer camps, but Connolly held the record for distance.

When there didn't happen to be a dance at Groveland, the gang often hit the road to Sonora, or Columbia, or even Angels Camp. In summer there were jigs at an open-air pavilion alongside the railroad track at Buck Meadows. For an occasional dance a group of railroad shopmen thought nothing of making the terrible drive to Yosemite, fifty miles further into the mountains over the narrow dirt road full of switchbacks and hairpin turns. Walt Magee drove them in his 1920 Olds — Bert Minard, the Schott brothers and Vince

Roundhouse and shop crewmen pose on the Hetchy's biggest locomotive, Shay No. 6, in front of the shop at Groveland, May 3, 1922. Through showing the picture to scores of people over the years, the author was able to get most of the men identified. Going from left to right, the top row from engine cab to smokestack: second man is Roy "Bud" Reese, Jim "Dad" Rasmussen (4), Buck Schott (5), Bert "Bud" Holmes (6), Frank "Butch" De Ferrari (7), Edmund "Ed" Coyle (8), Horace Hawkins (9), Vince Lawrence (12). Seated by dome: Walt Magee and Frank Schott. Sitting on running board: Roy Colburn (2), Charles "Tight" De Ferrari (3), Earl Durgan (4). On the ground: Master Mechanic Milo Jubb, Will Blanchard, Blacksmith Tom Hope, Byrd Long, George Prewitt, Tom Lumsden, Jake Laveroni, Bert Minard, last unknown. In the cab is timekeeper Ed Pillette. *(–City of S.F.)*

Lawrence – and said they never worried about the road. "We didn't have to work the next day anyway." The next day being Sunday, there were church services for those who had the strength to get up. The Protestant church was just west of town; the Catholic was Mt. Carmel Church at Big Oak. It has been told that if the Chief was in town of a Sunday, *all* Catholics were at Mt. Carmel when O'Shaughnessy walked into the small hilltop structure for Mass.

Every night at Groveland there was a regular parade to the post office. Susie White remembered this well: "We never got the same thrill on Fifth Avenue as getting dressed up a bit and going down in the evening to get our mail." Next to the post office, the most popular place in town at night was Laveroni's ice-cream parlor, run by his daughter Bernice. About once a month a man from Sonora drove through and showed a movie in the dance hall which was always well attended. Sometimes fabulous Italian feeds were served in that old Odd Fellows Hall at Big Oak, a building left over from Gold Rush days.

Over forty years later Susie and Jack White fondly recalled those pleasant construction days in Tuolumne County. "It was a grand experience," she said, "and they were very happy days." To Jack, "Things were different in those times, and it's very hard to express. But I think the essence of life on the Hetch Hetchy Project was that no one tore anyone else down to get ahead." The men worked extremely hard at their jobs, then threw themselves wholeheartedly into the fun when off duty. Entertainment wasn't given to them in an easy chair; they made their own and enjoyed it all the more. "Besides," added Susie, "they were a special breed of men up there, above the average; something special about them all."

Priest's Hotel, operated by Dan Corcoran, stood at the top of Priest Grade on Big Oak Flat Road, four or five miles west of Groveland and only a mile from the big construction camp at Priest Portal.

Carl Inn was a popular stopping place on the way to Yosemite Valley. Travelers often stayed the night or spent a few days on their way to or from the national park. *(—Robert O. Richardson)*

When Groveland's Europa Hotel burned to the ground one night in the early '20s, flames leaped across the road and scorched the Groveland Hotel. *(—Moccasin Powerhouse collection)*

12

GROVELAND SWINGS!

Groveland had retained a lot of its old '49er atmosphere long after more favorably located settlements had quieted down. So the boomtown fervor and "the boys" whooping it up on Saturday night didn't particularly shock Poison Oaker natives. They rolled with the punch and several of them participated to the extent that they were to be found running gambling and drinking establishments. Most of them also kept a tolerant eye on the painted ladies, or soiled doves, around town. Catherine Cobden Haight said: "I'll bet that in the whole town of Groveland there were only about two public places that weren't at one time or another whorehouses – the post office and the church." And from the recollections of various old-timers, we can safely conclude that bootleg supply houses were almost as plentiful. Gambling was a little better organized; that is, it was in the hands of fewer people.

Here's how a San Francisco *Call & Post* reporter described Groveland in 1922:

> . . . this is some torrid little town. The townsfolk are a fine, cooperating lot, but there's a pack of mountain dew hereabout – jackass whiskey with the kick of dynamite – and all sorts of gambling games. The hard rock men come into town "fat" from camps up the line. The keepers of the blind tigers and the faro banks and the like in Groveland set themselves the laudable task of reducing the hard rock men – taking the "fat" off them. Often the process is complete in a few hours' time.

Groveland went through "what almost would equal a gold rush or mining strike in the early days," according to the Sonora *Banner* in 1920. It received the onslaught of thousands of men, "allowed them to feed, fight and pass on."

The boom in "torrid little" Groveland was taking place right at the start of our infamous Prohibition era, from 1920 on. The entire country was in a turmoil; newspapers everywhere reeked with lurid tales of bootleg raids, and Tuolumne County was no exception. The Sonora *Union Democrat* in the first years of the 1920s reports frequent raids by "prohi agents" and the sheriff's men at such places as Jacksonville and Groveland, obtaining such incriminating evidence as peach brandy and red wine. Groveland, fortunately, was a long journey from the county seat at Sonora and arrangements could usually be made for the sheriff or his deputy to find something – but not too much. So the monthly fine became a sort of license, guaranteeing that the payer would not be troubled for another thirty days. That's how it went most of the time for approved outlets; amateurs and others were on their own. It would be an understatement to say that prohibition laws were not really any trouble at all to people on the Hetch Hetchy Project who needed a drink. There was plenty of hooch around – of all degrees of goodness or badness – at all times and in plenty of places.

Ray Carne, an old railroader on the Hetchy, wrote about a typical operation at a tiny settlement along the line: "Old John Fry who had a cabin east of the siding, did a little bootlegging. [He was a] good bootlegger . . . had an annual pass as far as the dam." Lou McAtee said he could always get a case of good Scotch whisky for parties, but some of the guests preferred bootleg. Lou knew a farmer at Oakdale who had various grades and ages of bootleg whisky buried in his orchard. "How good do you want?" he'd ask, pacing off so many steps from trees or buildings. Red Wanderer remembered this, too; he said it was $1.50 a gallon for the good stuff. Harry Chase, one of the best tunnel men in the business, remembered that in the early '20s nearly every family was making fig wine. "It had a real good kick, if you added a few potatoes."

Millard Merrill, who was elected Justice of the Peace at Groveland in 1924 and was later a County Supervisor for many years, said jackass whisky was being made out in the woods all around Groveland and neighboring Big Oak Flat and Second Garrotte. According to some old-timers, when demand got especially high, they'd get delivery warm from the still, but no one was able to recall a time when booze was not available. You could always plunk down a four-bit piece in several places in town and get a good shot of quality jackass or real stuff run in from Stockton. Sal Ferretti, the butcher, had a still on his ranch below the ball park; his sales were made at the slaughterhouse, according to Walt Magee.

Hetch Hetchy officials waged a constant battle to keep booze out of the camps. The cooks at Early Intake actually had a rig on back of the stove in which dried apricots or prunes were "converted," Doug Mirk noted. City officials hired Charlie Baird as a special agent to patrol their camps and surrounding areas, to keep the hooch out of camp and keep suppliers on the run. Baird had operated local stage lines for years and built and ran Baird Hotel in town. As special agent he was quick, firm and efficient, yet did his work in such a way that most of the men he policed remained his friends in later years. It wasn't an easy job when tunnel men would pour jackass into their "snoose" (chewing tobacco) and get loaded just chawing away on the job.

Jimmie Graham told about a typical incident while he was electrician at Second Garrotte. An off-duty miner one day stumbled onto a bootlegger's hideout in the woods and sneaked jugs of the stuff into his bunkhouse. Drinking out of empty fruit cans, he and everyone else on the shift got crazy on the booze; by 2:00 p.m. they were all drunk and unable to report for work. Then the day shift came out at 3:00 and joined the party! Charlie Baird was called and arrived from Groveland around 6:00 with a couple of trucks. "Charlie was a terror, especially when there was booze in the camps," said Graham. The drunks were put into the trucks, hauled to Sonora and paid off on the spot.

The San Francisco people didn't want any drinking in camp, but they rather smiled at what went on in town. When a good man got drunk and lost his money right there in the area, he'd be back on the job Monday and not away off in Stockton or the City looking for work. Baird told the author that after payday drunks would be lying all over the streets in Groveland. This usually happened four times a month: on two paydays and two shift changes. Some of the men in distant camps would work six or seven months without a break, then come to town with over $1000 in hand; it might last nine or ten days.

While the rough and well-heeled miners were getting loaded, they were given every opportunity to indulge in gambling. The infamous "Bucket of Blood" attracted a lively crowd, as did Shorty Nolan's various locations. There was another joint over the hill in Big Oak. But king of them all, the gambling joint with the biggest play in recent Mother Lode history, was "Peach" Pechart's well-organized operation at the Groveland Hotel.

Walter A. Pechart had relatives near Bower Cave on the old Coulterville-Yosemite Road. He returned to Groveland with the start of Hetch Hetchy construction, leased the historic hotel from Tim Carlin and set to work at once putting in games, hiring bartenders and converting some rooms at the back into a small "bull pen." Any number of people who were there have told how the off-duty workmen would hit town and "home in" on the Groveland Hotel, full paychecks in hand, ready for a night of fun — maybe even two nights, if unusually lucky. Customarily, by next morning the money would be all gone, some spent and the balance lifted without a struggle.

When Millard Merrill was Justice of the Peace, a Hetch Hetchy worker came to his office one morning and complained that Pechart had rolled him and taken his check. Merrill just called Peach in and asked him to return the man's eighty bucks, which he promptly did, peeling the paycheck from a roll of them "as big as a smokestack." On the very next day the same fellow went into Pechart's,

The railroad station at Groveland sat just across the tracks from Hetch Hetchy hospital. To reach town, passengers went to the right of the picture, crossed a footbridge over Garrotte Creek and ended up at the east end of Main Street. Track buses Nos. 21 and 24 are headed west, toward Hetch Hetchy Junction. *(—White Motor Company)*

cashed the check and spent all the money. "Pechart was a corker," laughed Merrill.

Many tales are told about what went on at the Groveland Hotel in the process of cleaning out the tunnel stiffs and sending them back to the job. A lot of people disliked Peach and his methods, but there were others who thought he was a pretty good guy. George Laveroni was one of these; it was his opinion and that of most other Groveland natives that Pechart made his first million there during Hetch Hetchy years. Peach reportedly had a couple of semi-silent partners, "but I get the first cut," he told Laveroni. Possibly the key to Pechart's success at Groveland was diversification: gambling, drinking and prostitution available in one handy location.

Although the Groveland Hotel's bull pen is best remembered for its girls, it was only one of many "sporting houses" in full-steam operation. At all places the standard price was three dollars, better part of a day's pay at that time. Phil Hope, who was a teen-ager at the time, said there were sometimes as many as six girls operating at Priest; they circulated around the bar, then took customers to their tents out in back. Hope remembered that a woman was working at the Charlotte Hotel in Groveland and two upstairs at the Iron Door. There was also the red-light district at "Pecker Point," just a short walk east from the railroad shops. Here were two or three shacks, one of which burned one busy night, seriously injuring an inmate. Ernie Beck, who at one time worked on the railroad as locomotive hostler and call boy, had three train crews to wake up each morning starting at 4:00 a.m. He said he often saved himself a long walk up through town by checking at Pecker Point first to see how many of the trainmen might be there sleeping it off.

Another popular place, according to George Laveroni, was a bawdyhouse down on the old Watson Ranch. It was run by "Jimmy the Wop," a fellow who could really play the piano. Farther to the east, just outside Big Creek camp, was a big house, always busy. Then at Buck Meadows a couple of squaws from Coulterville set themselves up in a cabin and had a steady stream of callers from three Hetch Hetchy tunnel camps: South Fork, 5-6 and 8-9. Occasionally the country houses had troubles, as witness a 1922 report in the *Union Democrat* of Sonora: three masked men robbed a "resort conducted by a woman above Groveland last Saturday night." Three women and four or five men were there at the time and the ruffians "cowardly shot one of the women." Some of the distant camps didn't have regular houses, but on paydays it was not uncommon to find a couple of the painted ladies performing in tents just outside the areas. Apparently these girls of the "floating" gang were transported by their pimps ("P-Is") to wherever there was promise of money or action.

Not every Hetch Hetchy worker patronized the red lights, by any means. While the great majority weren't married, some were courting steady girl friends and there were plenty of Lovers' Lanes to enjoy with Becky, Mabel, Alice or whomever. A sober man could always borrow a pal's car for the occasional heavy date. It has even been reported that on more than one occasion a railroad track bus might be observed "parked" in some lonely mountain siding far from roads and civilization.

Well, the miner, logger, powder man or dinky skinner, after he'd finished with the amusements and given up trying to beat the gamblers, would most likely be found with his buddies sleeping it off along Main Street.

Hetch Hetchy's Heisler No. 2 and newly acquired Mikado No. 3 get steam up in the early morning sunlight outside the shop building at Groveland, May 10, 1919. *(—City of S.F.)*

In 1921 the railroad acquired a large three-truck Shay, No. 6. It was so effective in dragging heavy cement trains up the Priest Hill grade that another big Shay was rented late in 1922 from Northwestern Pacific while HHRR No. 4 was being rebuilt after an accident. *(—Marvin T. Maynard)*

13

RAIL LINES ALL OVER THE PLACE

As crews had pressed ahead with heavy construction work on O'Shaughnessy Dam and the aqueduct, and work started on a reservoir at Priest and the powerhouse at Moccasin, railroad operations moved into full swing. There were still men and supplies to be hauled, but now sight-seers and official parties were using the line, and tons of cement were needed daily at the dams and tunnel. Rail lines, inclines, cableways and spurs in various gauges ran all over the place to connect the main line with the busy work sites.

The Hetch Hetchy Railroad carried on its operations in steam with the two Heislers and the rented Shays until January 1919, when its first rod engine, a Baldwin-built 2-8-2, arrived at Groveland. Bought secondhand in Ohio and assigned No. 3, she wasn't really satisfactory, as a look at the diary kept at headquarters will demonstrate. According to the diary, No. 3 arrived on the 23rd and was sent out on her first trial run on the 25th. "Trial successful as to curves, but pulling was not in excess of three cars," reads the entry. Six days later, the bad news: "No. 3 engine developed boiler trouble." Three months later, on the first of May, the record shows this engine returned from the Southern Pacific shops at Sacramento after heavy and expensive boiler repairs. The cost to overhaul the 3-spot and convert her to oil was $9000.

Whoever was responsible for buying this engine was certainly not on the ball. She spread rails on the curve at Rattlesnake and Six Bit and went on the ground – her rigid wheelbase was too long! The tires on her front drivers slipped loose on the hill at Cavagnero and the whole train was hauled back to Hetch Hetchy Junction by No. 2. All this happened within a month of her return from major shop work. The management had an interesting and undoubtedly unique solution to the track spreading she caused. The office diary records that on May 6, 1919, additional bunkhouses would be built at Priest "for track work maintenance for No. 3." Yet, someone was happy to have the big rod engine at work (or maybe they thought she wouldn't be around long), for on May 10 it was arranged for the City's photographer, Horace Chaffee, to photograph the Mikado with a train. The pictures were up to Chaffee's usual fine quality, for which historians and rail enthusiasts are ever grateful.

Three steam locomotives were bought new by Hetch Hetchy officials after their sad experience with the 3-spot. No. 4, the favorite, was an attractive 95-ton 2-8-2 built in January 1920 at the American Locomotive Co.'s Schenectady Works. No. 5, a nicely designed 2-6-2 Prairie type, was turned out by the Cooke Works in March 1921 and was put into passenger and mixed-train service. Finally came No. 6, the biggest of all: a huge 101-ton, three-truck Shay built by Lima. This engine went to work in the spring of 1922 on the slow haul of cement trains up Priest Hill, where she succeeded admirably. In fact, later the same year a similar large Shay was rented for a couple of months from the Northwestern Pacific while No. 4 underwent extensive repairs as the result of a bad accident.

Several retired HHRR employees mentioned that a third big Shay, the Sierra's No. 12, was also used on the Hetch Hetchy from time to time. Walt Magee, who had worked on No. 12 at James-

Engine No. 3, from a railroad in Ohio, is posed with some of the shopmen on the Groveland turntable, May 10, 1919. The first three men at upper left are Frank "Butch" De Ferrari, George Prewitt, and Master Mechanic Milo Jubb. Standing on the pilot beam is Herman Kirchen, while Tom Lumsden is shown at far right, leaning on the coupler.

Later the same day No. 3 set out for the junction with a freight train, taking City Photographer Chaffee along to record the 3-spot's performance on film. The photo below shows this train leaving Groveland, climbing the grade about a half-mile west of town. *(—Both, City of S.F.)*

Builders' photos of the last three steam locomotives acquired by HHRR, all bought new after the expensive and unhappy experience with secondhand No. 3. Above, No. 4, a 2-8-2, was turned out by the Schenectady works of American Locomotive Co. in 1920. *(–Adolf Gutohrlein)*

No. 5 was a Prairie type, 2-6-2, intended for passenger and mixed-train service. Also built by American Locomotive, it was turned out by Cooke works in 1921. *(–Vernon J. Sappers)*

The largest HHRR locomotive was No. 6. Weighing nearly 100 tons, it was built by Lima Locomotive Works in 1921. Designed for low-speed, heavy-pull operation, the Shay worked well in cement train service on long, steep Priest Hill.

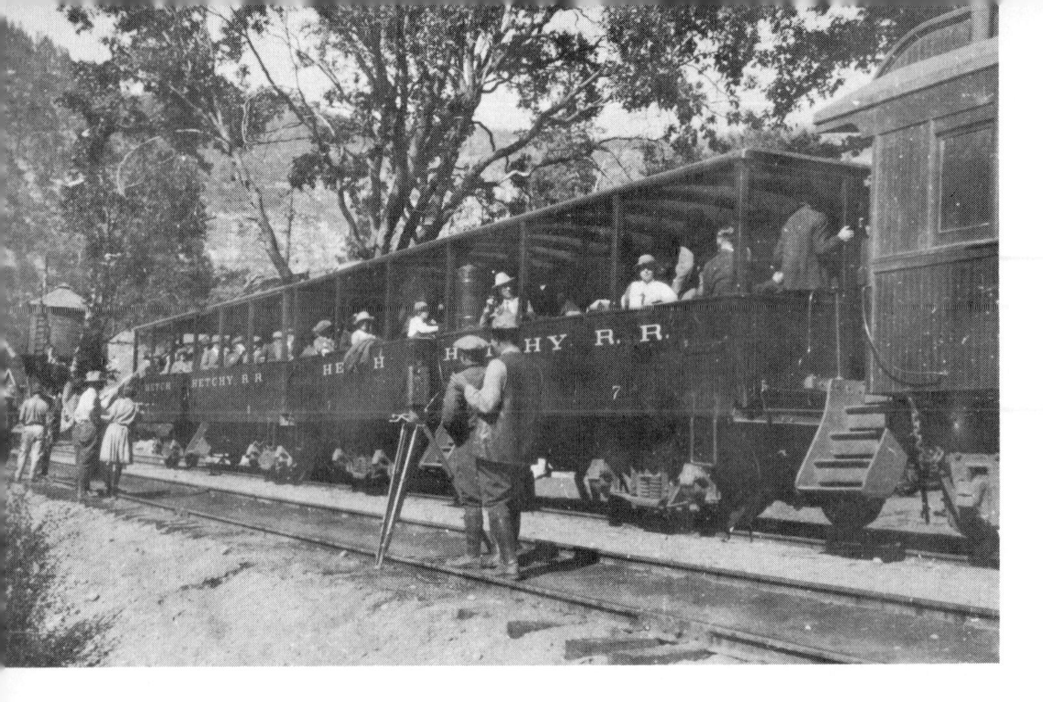

Two of Hetch Hetchy's five passenger-train cars were these open-side "rubberneck" excursion cars, Nos. 6 and 7. They were built at the car shop in Groveland on the underframes of two old gondola cars. Shown here at Hetch Hetchy in 1922, they carry a special party from the American Society of Civil Engineers.

town and knew her well, said that in the early days before ballasting was completed this engine worked exclusively between Mather and Damsite. In 1924 she was rented again and used to haul trains of penstock pipe on the steep spur to Moccasin Powerhouse. These were special operations, of course, and for most other work the Mikes proved themselves superior, even No. 3 after the Hetch Hetchy had finished rebuilding her. General Manager Al Cleary in a 1921 talk described the 2-8-2s as giving no trouble whatever, being much faster with equal loads and requiring much less upkeep than geared power.

One of the busiest periods of railroad operation was 1922, for the big dam was completed early the following year and tourists and official parties were coming to inspect the work. At the peak the railroad had five pieces of passenger equipment: a 58-foot combination baggage and passenger car, two 45-foot coaches, and two "excursion observation coaches" of 39½ feet, the latter built at Groveland on the underframes of a couple of excess gondolas. Also on the list of rolling stock were 14 flatcars, 12 boxcars, 14 assorted gondola and dump cars and 3 cabooses. Service equipment included a steam shovel, a tank car to carry oil and water for it, and a snowplow.

During the first year of operation, locomotives had been repaired in an open-air "shop" alongside the main line, but by 1919 everything was under cover in a building measuring 50 x 155 feet, shop and enginehouse combined. It contained machine and blacksmith shops and had two tracks running through. The crew here could do practically all locomotive overhaul and repair work, as well as heavy repairs to construction equipment. Heavy boiler work and forging were sent to Sacramento.

The majority of the Groveland shop personnel were Poison Oakers or from adjoining Mariposa County, like blacksmith Tom Hope. The master mechanic was Milo Jubb, who came from the Yosemite Lumber Co.; he and his assistant, George Prewitt, supervised a crew of 45 men. By all reports, Jubb was not popular and the men were constantly harassing him with such antics as removing the nails in the seat of a new outhouse so that Jubb would fall in. "Dad" Rasmussen was in charge of the car-repair shop, alongside the main building, and under his supervision some rugged rolling stock was turned out for the railroad. Harry Stenander was boss of the shop for gasoline-powered vehicles.

Big Tom Hope, the blacksmith, was a man liked by everyone and had a sense of humor that was hard to match. When someone asked him to have a look at his inoperative dollar watch, Hope "fixed" it with an air hammer! Another time, after a bad session with Dr. Degnan, Hope made a special surgical knife for Doc, with a big wooden handle so it wouldn't get lost inside a patient.

"Mac" McAfee was Superintendent of the railroad as well as being Construction Engineer in charge of all Mountain Division work. Everybody

Engine No. 6 stands outside Groveland enginehouse-shop, early on a summer morning in 1922, ready for another long day's work dragging trains of cement up Priest Hill. The Shay's side cylinders, connections and gears, through which power was transmitted to the wheels, are all on the right-hand side. Consequently, the boiler on this type of locomotive was offset to the left in order to balance the weight. *(—Photo by Ernie Beck)*

Mikado No. 3 at Groveland. Superintendent Walter Dresser of the San Joaquin & Eastern R.R. paid a visit to HHRR and other mountain rail operations in the early '20s to check experience with various types of motive power. He is shown here beside the 3-spot. *(—Dresser collection)*

No. 4 with mixed eastbound train makes a winter stop at Smith Station to unload merchandise and a couple of passengers for the near-by tunnel camp at Adit 8-9. *(—Roy Brooks)*

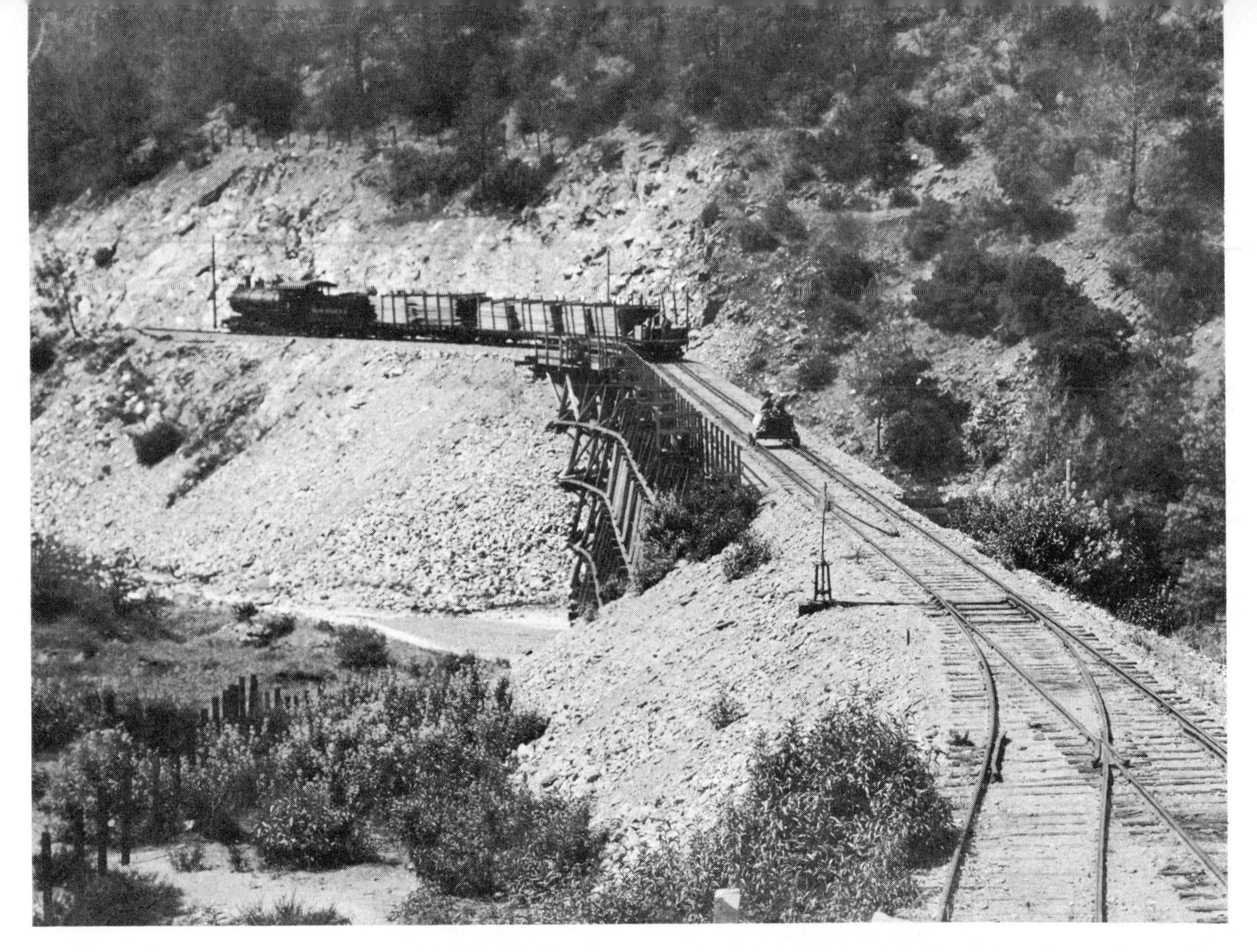

Six Bit Gulch in 1919. Photo was taken from top of caboose in siding as a westbound train of lumber with engine No. 3 stops across the old timber trestle. Tracks were alongside the Tuolumne River at this location; the grade was later raised and a higher bridge constructed when waters of Don Pedro Reservoir backed up into the canyon.

liked Mac: he had a wonderful personality to go with his engineering ability. The McAfee family lived upstairs in the City's headquarters building, while engineering and accounting offices filled the lower floor. Willis O'Brien was Chief Accountant, assisted by Bob Scott, Fred Arndt and Clara Newport. Marie Floyd Neel was phone operator, and Lincoln Wilson and E. J. Case also worked in the office. This gang ran many of the Groveland dances, picnics, even "snowball" special trains up to Hetch Hetchy. The men in the near-by railroad shop always joined in any parties, especially Ray and Walt Magee.

Another well-liked and respected official was Julian Harwood. He was a fine construction engineer with railroad experience but preferred to work as Roadmaster on the railroad. Every Christmas Harwood personally delivered turkeys to the various camps on his track speeder.

The central warehouse for the Hetchy's entire Mountain Division was also at Groveland, almost alongside the railroad shop. Ed Erskine was in charge and many employees first went to work for the City in his department. Being a part of headquarters at milepost 27, the warehouse was ideally situated: it could send parts and supplies in both directions along the project, even in the stormiest weeks of winter, using the railroad as its highway.

Running in conjunction with the Hetch Hetchy there was an independent feeder railroad, the logging and lumber line of the California Peach Growers Inc. It made connection at Mather and served the fruit association's big two-band sawmill, a mile and a half south of the station. Logs were brought to this mill by extensive rail routes which varied from time to time, and finished lumber was loaded and handed over to the Hetch

Sawmill of the California Peach Growers Assn. was a mile south of Mather. This mill supplied HHRR with five or six outgoing carloads of lumber a day, during cutting season, from 1919 to 1925. Skid road and donkey engine are in foreground. *(—City of S.F.)*

Hetchy. As early as 1920 they sent out about five carloads of shook material per day.

The Peach Growers operated three geared steam locomotives at Mather: a decrepit-looking Type A Climax, a two-truck Shay and a new 1923-model Climax which came painted with the company's new name: California Peach & Fig Growers. These engines were apparently maintained at the Hetchy's Groveland shops on a contract basis. Walt Magee told about a speeder ride up to the Peach Growers mill to install some new flues on one of the locomotives. The sawmill's product, mainly material for fruit boxes, provided the City's railroad with freight revenue from the time it was opened as a common carrier until the last day of such operation — and then even beyond that, as we shall see.

When the Hetch Hetchy Railroad was in the planning stage, there were two other large lumber operations proposed for the area. However, White & Friant never did get started, and South Fork Lumber at Hardin Flat wasn't a heavy shipper.

All other rail operations affiliated with the Hetch Hetchy were part of the City's project. The earliest was a narrow gauge used at the Lake Eleanor Dam construction site in 1917-18. There side-dump cars were lowered by gravity from aggregate bunkers to a mixer, from which one-yard bottom-dump cars carried wet concrete to the pouring area. The latter were steam operated, the cars being pulled on an endless cable powered by a single-drum hoisting engine.

Two years later, in Hetch Hetchy Valley, a real steam narrow gauge, the Valley Railroad, began operation to service construction of O'Shaughnessy Dam. The two inclined cable railways here, one of them a combined standard and narrow gauge 700 feet long, added to the railroad's growing complexity.

Of course, the greatest incline of all was the 42-inch-gauge railway at Early Intake, dropping down the mountainside to the portal camp and powerhouse. At the bottom was a narrower-gauge system, the 24-inch tunnel railway using battery

The "Peach Growers Railroad" connected with HHRR at Mather, bringing in carloads of lumber from the California Peach Growers Assn. sawmill to the south. The main line was about 1½ miles long and several lines ran out from the mill into timber-cutting areas. The first locomotive was this Class A Climax type, built in 1911 and obtained secondhand from Seattle Car Mfg. Co. in 1918. Photo at Mather, May 1919. *(—City of S.F.)*

Peach Growers also purchased this Shay, No. 2, in 1918. It had originally been built for the Sugar Pine Railway, east of Sonora, in 1910. This 1922 photo was taken near the mill.

The Shay was sold in 1923 and replaced with this second No. 2, a brand-new Climax bearing the new name, California Peach & Fig Growers. The engine is at the mill in early 1924 during the winter shutdown. After HHRR ceased common-carrier operations in 1925 the Peach & Fig Growers operation was ended and No. 2 sold to S.J.&E. R.R. *(—Ferrill photo from Barney Emerson)*

HHRR station and water tank at Mather in 1922, looking eastward. The Peach Growers Railroad went off to the right and crossed over a low ridge to reach the millsite.

This load of sugar pine logs is being brought into the Peach Growers mill behind the Class A Climax, No. 1. *(—Ferrill photo from Barney Emerson)*

Below is the builder's official photograph of second No. 2. The Climax geared locomotives transmitted power to the wheels through a central drive shaft, driven by the inclined cylinders alongside the boiler. *(—D. S. Richter)*

Gravel Spur Siding near Buck Meadows supplied ballast material for Hetch Hetchy Railroad tracks, while City crews finished the trackwork that the contractor had left undone. The railroad's Marion steam shovel loads cars lettered for City & County of S.F., while No. 2 pops her safety valve on a nice May day in 1919. *(–City of S.F.)*

locomotives. Across the river, in post-construction years, another two-footer operated along the top of Lower Cherry Aqueduct. Powered by a Chevrolet gasoline locomotive, this line carried heavy sections of pipe to replace the flume.

Westward along the Hetch Hetchy in the vicinity of Buck Meadows, three small inclined railways had been designed. Two were to serve tunnel adits 5-6 and 8-9, but rough roads were used instead. The third was to assist in obtaining ballast materials at the "Gravel Spur" siding.

At Big Creek Shaft the railroad laid a short spur right up to the headframe during tunnel concreting, to facilitate lowering of materials. Earlier there had been a narrow-gauge, mule-powered tram at the surface to carry away the spoil.

On a wooded hillside above Priest, western terminus of the Mountain Division Tunnel, was the Hetch Hetchy main line and from it two cable inclines ran down, one a counterbalance line to the tunnel portal and the other serving Priest Dam construction work, down the canyon of Rattlesnake Creek. On the opposite side of the canyon a pair of twenty-ton Porter 0-4-OT steam locomotives worked with four-yard dump cars around the clock, carrying material for the earth-fill dam. This system was three-foot gauge.

The Hetch Hetchy Railroad had one branch, a steep, 7000-foot-spur serving Moccasin and the powerhouse site. It came downhill from the main line on a constant 5% grade, heaviest on the system, and actually connected with a standard-gauge inclined cable tramway that went right back up the face of Priest Hill alongside the penstock line.

Across the powerhouse's little valley, where the Foothill Division Tunnel would start, there was later a small two-foot-gauge system serving an aggregates plant. Battery tunnel locomotives provided the power, but they had cable assistance on the steep upgrade to the level of the old highway. Cable and electric combined, somewhat like the Fillmore Hill car line in San Francisco, but this affair was strictly one-man operated – including the cable!

More rail operations would be evolved as the Foothill Division work proceeded westward. One of the most interesting setups was to be found five miles west of Moccasin, at Red Mountain Bar.

The inclined cable railway known as "Intake Tram" would certainly qualify as one of the busiest of HHRR connecting lines. Originally a three-rail semi-counterbalance line, it was later changed to a single line with just the one car, shown here with a party of City Supervisors and their families on the week end of O'Shaughnessy Dam dedication in 1923. *(—City of S.F.)*

Here the Tuolumne River Canyon would be filled by the waters of Don Pedro Reservoir, and the aqueduct was designed to cross underneath by means of a siphon. The railroad came twisting out onto the canyon face here, about six miles out of Hetch Hetchy Junction. A spur ran left to the Pedro Adit tunnel camp, and a cable incline dropped down to the right, serving the siphon construction work in the riverbed. From a station-type platform beside the main line a Lidgerwood overhead cableway 2300 feet long carried freight and passengers across the canyon to a platform by a waiting narrow-gauge train. This was Brown Adit Tramway, a little over a mile and a half of 24-inch-gauge track with a six-ton Plymouth gas locomotive. It had direct connection with the tunnel railway, forming a system actually over 17 miles in length.

Finally, at the Foothill aqueduct shaft at Hetch Hetchy Junction there was a small Plymouth gas locomotive to work on the spoil dump immediately north of the yard tracks. A spur was later laid in from the Sierra Railway to pick up loads of tunnel granite for ballasting.

The City was not able to use railroad facilities in its logging operations at either Canyon Ranch or Mather, or we'd have two more systems to consider. These sawmills were both in restricted areas involving fairly steep hillsides, so donkey engines and skid roads were employed. But the railroad carried away the finished lumber, supplying various construction sites with forms, bunkhouses, bridges, ties, station buildings, whatever.

It can be said without fear of contradiction that the Hetch Hetchy, with its various "connecting lines," was one of the most interesting railroad operations ever devised in the short space of only 68 miles. Anyone who would undertake to model it in its entirety would have to be something of a genius and highly dedicated.

Two of the unusual passenger railcars operated by Hetch Hetchy Railroad, where they were called track buses. The Hetchy had a quantity and variety of gasoline-propelled vehicles not equaled by any other short line in this country; in fact, few large railroads could compete in variety. No. 20 (above) had been a highway and sight-seeing bus in Yosemite (note front fenders left over from road vehicle days), while No. 24 was designed just for HHRR operation. Driver Ivan Harten has stopped No. 20 for a picture in the canyon; No. 24 is ready to leave Mather as Train No. 3, on an early afternoon in March 1923. *(—Bottom, City of S.F.)*

14
GASOLINE ALLEY

The Hetch Hetchy Railroad was properly renowned for the number and diversity of gasoline-engined vehicles running on its rails over the years. The line was featured in scores of magazine articles, both in the engineering field and in general circulation. Perhaps the fact of municipal ownership and operation was responsible for this diversity; certainly no privately owned railroad used such a variety of internal combustion power over such a long period of time. In this service there were bound to be frequent calls for special train runs, for emergency delivery of equipment or for ambulance service. Road vehicles were useless in winter and turning out a steam train for each run would be time-consuming and expensive.

First came the converted Cadillac inspection auto of 1917, followed by the four motor trucks fitted with flanged wheels at Municipal Ry. shops in the City. The first true railcar seen on the Hetch Hetchy, Meister-built No. 61 of the Ocean Shore Railroad, put on a test and demonstration run in February 1918. So the Hetchy went to A. Meister & Sons of Sacramento a year later and ordered a car built on a White truck chassis, with undercarriage work by the San Francisco Municipal Ry. This was "track bus" No. 19; it could carry 13 passengers but was primarily outfitted as an ambulance car. No. 19 turned out to be the favorite of Chief O'Shaughnessy, who always had it assigned when he was up from the City on business or inspections. Earl Hope remembered that No. 19 was at first decorated with red and black paint, with gold lettering and numbers and ornamental tassels. She was speedy as well as pretty – old employees said she was fastest of all and often hit 50 mph running in overdrive on the straight stretches at Buck Meadows and Smith. The Chief liked this car's economy, too, for in a report dated as late as 1931 he claimed that she got 16 miles to the gallon.

No. 19 set the trend and there followed five more track buses, all Whites (more or less) and no two alike. During 1920 and 1921 two were bought new and two others secondhand. No. 20 came with its driver, Frank Orth, from the Desmond Park Service at Camp Curry, Yosemite; railroad wheels were applied at Groveland. Nos. 21 and 22 came new, but with completely different bodies built onto the White chassis by Meister: No. 21 to carry 27 passengers and No. 22 for mail, express and small freight. No. 23 had been a narrow-gauge railcar on the Nevada-California-Oregon Ry. Upon arrival at Groveland, she was split down the middle, widened, given a big Waukesha T-head engine and put into passenger service. She carried 24 people and was used mainly with No. 20 on park-operated trips from Mather to Damsite, so that Yosemite tourists could view the construction work in one- and two-day trips from the valley.

The final track bus was No. 24, most unusual of all. Express bus No. 22 was badly damaged in a head-on collision with engine No. 3 in 1922. Needing more passenger capacity, the railroad scrapped what was left of the superstructure and built a novel chassis at Groveland. Behind the rear axle was mounted the engine, enclosed in a cage which stuck out behind the car's body. As a result, a very low-hung body could be designed, one with great seating capacity, a low center of gravity and minimum vibration, noise and engine

The Cadillac touring car was first of HHRR's long line of track buses and trucks. It is shown being overhauled at Groveland, early 1919. Reportedly the last use of this vehicle was to carry officials and their guests on fishing trips into the mountains. *(—Louis Stein, Jr.)*

Track truck No. 6, a Packard, is running as a freight extra (note white flags) east of Intake station. The time is 1919 and the unique "train" is headed for Damsite. *(—Vince Martini)*

Speeder No. 16 on a passenger run! It was the day after Christmas, 1919, and Nels Eckart, engineer in charge of Mountain Division construction, was taking Mrs. Eckart and three visitors from the East for a look at the project. Mrs. Eckart remembered that they had put sheets of newspaper under their overcoats to keep out the biting wind. *(—Mrs. Grace Eckart)*

No. 19 served for many years as ambulance during Mountain Division construction, carrying injured (and dead) to the hospital at Groveland. With its self-contained turntable under the floor, No. 19 could be jacked up and turned anywhere along the railroad. Harry Stenander, in charge of gasoline power, demonstrates operation of the table. *(—White Motor Co.)*

No. 20 sits discarded at the track bus roundhouse at Hetch Hetchy Junction in 1937. In background are cars of cement on Sierra Railway, to be hauled to Mather during dam-raising project.

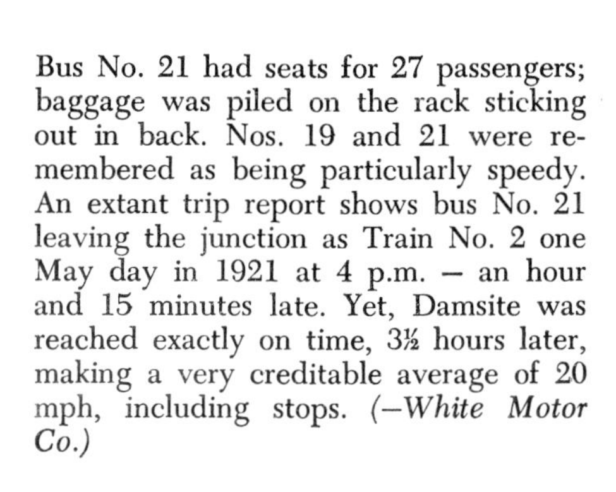

Bus No. 21 had seats for 27 passengers; baggage was piled on the rack sticking out in back. Nos. 19 and 21 were remembered as being particularly speedy. An extant trip report shows bus No. 21 leaving the junction as Train No. 2 one May day in 1921 at 4 p.m. — an hour and 15 minutes late. Yet, Damsite was reached exactly on time, 3½ hours later, making a very creditable average of 20 mph, including stops. *(—White Motor Co.)*

No. 22 was the only one of its kind: a gasoline-driven rail-car for baggage, express, mail and small freight shipments. Designed by Meister & Sons of Sacramento, the body was mounted on a standard White chassis having cowcatcher and railroad wheels already in place. These photos by Meister show various details of the car. Below, driver's compartment had simple controls and a "steering wheel" converted to operate the brakes. The car had a short life, being wrecked in a collision in 1922.

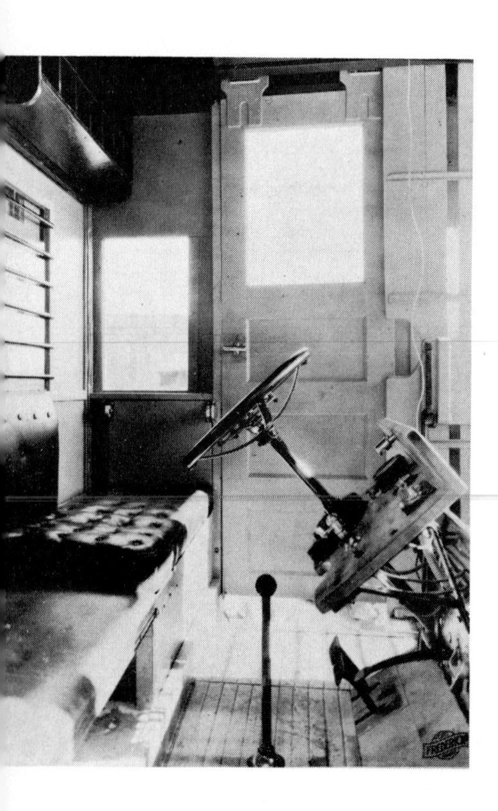

smoke. The completed chassis was then shipped off to Sacramento where Meister's experts applied a de luxe closed body. *Railway Age* reported that No. 24, weighing only 10,400 pounds, "is probably the lightest in proportion to its seating capacity (32) that has yet been built."

Named "The Freak" by her operators, this new track bus was designed to climb the twisting, twelve-mile Priest Hill grade at a steady 27 mph and would average eight miles per gallon of gas. But first there were some problems. The car couldn't be made to lubricate properly and burned out the bearings two or three times; sometimes the weight at the rear caused a bad shimmy effect. Occasionally gear shifting could be embarrassing, for the long connections would become disengaged and that would be that for the day. This happened, it was discovered, when a capacity load caused the floor to sag – and there were often full loads, with 1800 to 2000 men working on various parts of the project needing transportation. This trouble and the others were soon corrected and No. 24 had a very long, active life on Hetch Hetchy Railroad.

The Hetchy was equipped with Sheffield track speeders at first, then more were added from Buda for a total of twelve. These perhaps were numbered from 2 to 5 and 11 to 18, for those are the missing numbers in the gasoline roster, with No. 1 being the old Cadillac, 6 to 10 the track trucks, and 19 to 24 the buses. No. 18 appears on train sheets as a speeder in track maintenance work, so this assumption may be valid. This unusual railroad, of course, had an excess number of speeders because of the innovative use made of these light, economical vehicles.

As more elaborate railcar-type vehicles appeared in 1920 and 1921, some of the earlier ones were discarded or changed. One of the Pierce-Arrow trucks was outfitted to serve as a fire-fighter truck on rails – the Hetchy's fire train – and the other was made into a sort of motorized tank car, carrying fuel oil from Hetch Hetchy Junction to a storage tank at Cavagnero siding, above Moccasin. The worn-out Cad was relegated to occasional fishing trips, then discarded. Other trucks were changed back to ordinary road vehicles. Eventually, when passenger service dwindled to nothing in the mid-1920s, track bus No. 21 became a track truck with stake body. Nos. 20 and 23 were stored, though used rarely on inspection trips. Nos. 19 and 24 lived on actively as summer and winter line cars. In later years, with complete cessation of steam train service, the railroad began another phase of gasoline power, operating three Plymouth locomotives, a rotary snowplow and other interesting pieces.

It might well be asked at this point: How did Hetch Hetchy operate a common carrier railroad with all this light stuff running around, some of it even on personal matters? For one thing, there was an elaborate telephone system connecting all stations, sidings and spurs with the dispatcher's office at Groveland. Each movement was dispatched and shown on a regulation train sheet, so the situation was well controlled, though it might appear to have been otherwise. Bus No. 19 was usually on call as the ambulance, dispatched as an extra train to accident locations. On many trips, according to Ray Carne, both stretchers were occupied and Doc Degnan was right at work in the car. While buses 20 and 23 took care of sightseeing runs between Mather and Damsite, Nos. 21 and 24 held down mainline train runs, at first just east of Groveland, then later over the whole line as steam passenger trains were shelved. Employees' timetable No. 6, effective May 1, 1921, shows "motor" trains only: Trains Nos. 2, 4, 6 and 12 eastbound, and Trains Nos. 1, 3, 5 and 11 westbound. Nos. 11 and 12 were "freight" trains and all trains operated daily. Checking old copies of Form 38, "Engineer's Trip Report," for the same month in 1921, it is noted that track buses 21 and 19 were on passenger runs one day and express-LCL bus 22 was the "freight."

Each driver put in an approximate eight-hour shift by ingenious scheduling over three-day cycles, operating both passenger and freight one-man trains. For example, a driver would be scheduled to leave Damsite at 3:00 a.m. as Train No. 1. (Remember, the poor devil had to crank the "train" by hand; there were no self-starters. Imagine the scene in January!) He arrived at Hetch Hetchy Junction at 8:20, went back as far as Groveland as Train No. 12 (9:00 to 11:30) and was off duty. The next day this same man would reverse, taking

Track bus No. 23 came to the Hetch Hetchy secondhand, having served previously on the narrow-gauge Nevada-California-Oregon Ry. She is shown above unloading passengers at Oakland Camp after an excursion run to Damsite and back in 1922. To the right is No. 24, signed as carrying the mail, ready to highball out of Mather early on a February afternoon in 1923. At the throttle is Ray Carne. It was on this run that popular Ray often carried sacks of freshly caught trout back to family homes in Groveland, according to Ernie De Ferrari. No. 24 had her engine cantilevered out to the rear, as shown below, allowing the big 32-passenger capacity and a lower center of gravity for speed on the many curves. *(—Above, Robert O. Richardson; right, City of S.F.; below, builder photo by Meister)*

freight No. 11 from Groveland at 12:45, arriving at the junction at 2:25, then all the way back to Damsite as No. 2 with a different motor, arriving at the end of his run at 7:50. The third day would be spent taking Train No. 11 from Damsite to Groveland and going back as No. 12, approximately eight and one-half hours on duty. Thus, only three men kept four scheduled through trains operating over the system each day. Trains Nos. 3, 4, 5 and 6 were locals with shorter runs, the round trip taking only one working shift. Heavy steam freight trains, backbone of the railroad operation, were all run as "extras" around the clock with no set schedules, hauling bulk cement up and forest products and whatever else back.

There was one drawback noted on timetable No. 6: "Owing to limited space, the City does not guarantee immediate transportation of baggage on motor cars." But the railroad gave plenty of service and kept costs way down. Anyone who studies the Hetch Hetchy Project can come to no other conclusion than that Chief O'Shaughnessy was for the most part a careful manager of San Francisco's limited funds. There were some who thought him too careful. Ray Carne recalled a special off-season party trip with bus No. 23, picking up San Francisco's Supervisors at the junction. When it came to tipping the driver the Chief announced, "Fifty cents is enough, boys," much to Carne's disgust. Ernie Beck remembers this party arriving at the Groveland shop area in that "rubber neck wagon" and in the depths of Prohibition. The politicians jumped out, breathed the fresh, piney, mountain air and gave vent to their usual cheer: "Hetch Hetchy . . . where's the hooch?"

Track buses had all-wheel brakes operated by turning the steering wheel. This could be very tricky, according to Bob Cheminant, son of the Trainmaster; sometimes the brakes would tighten or release as the vehicles manipulated sharp curves. Several operators deliberately frightened greenhorn passengers by looking around at the scenery and taking their hands off the wheel at cliffside curves. Men working on the Hetchy job never got tired telling about it in later years: the fine people they worked with, the fun they had. Fun? Like the time a dozen or so track torpedoes were exploded when No. 19 arrived at Damsite with newlyweds. Sure was funny when the car was derailed, but not when out stepped one of the passengers, McAfee, the Super himself!

Les Phelan, who was working on and around the railroad right up to its end in 1950, recalled perhaps the most unusual run of all for the track buses. When the hospital at Groveland was destroyed by fire on July 27, 1922, and nurse Ethel Moyer was badly burned and injured, it was No. 19 that made an emergency run with her down to Hetch Hetchy Junction, then up the Sierra Railway to Sonora, where she died the next day.

Earl "Cappy" Caplinger once took the whole baseball team from Groveland over to Tuolumne City for a big game, using a Sierra Railway pilot from the junction. Bus No. 23 was the transportation and, remembering how hard it was to crank that big Waukesha motor, Earl backed her around the wye at Tuolumne upon arrival so she was facing west, ready for the return. After the game, the players gave him a little shove and she started under compression with a minimum of trouble. Ray Carne remembered taking the team to Sonora for another game, while Walt Magee said Hetch Hetchy track buses even went as far as Angels Camp for one ball game, over that remarkable Sierra Railway branch line.

Indeed, the Hetch Hetchy Railroad had a unique operation going up in the Sierra Nevada during the early 1920s. There may possibly have been other railroads with a similar variety of equipment, but not such a short railroad, and certainly not operating steam trains and a flock of railcars, speeders and other gasoline power in such a way that the single-track line performed like a fairly busy highway.

Train No. 6 at the Groveland "passenger depot at Hospital Road Crossing." On this particular morning it appears to be "double-headed" with buses Nos. 21 (right) and 24. *(—White Motor Co.)*

Operating the Hetchy in wintertime was always a special problem, as snows are heavy above elevation 4000. Engine No. 4 pulls her mixed train eastbound at Smith, shoving wedge snowplow No. 101 ahead to clear the rails. *(—Roy Brooks)*

At the summit, Poopenaut Pass, 5064 feet above sea level, winter snows greeted HHRR crews for several months each year. Snowplow No. 101 is clearing out the cut three miles east of Mather. *(—Mary Male)*

Plenty of cool water for locomotive boilers at Damsite in 1922! During dam construction in winter months, the concrete was mixed with warm water. *(—Mary Male)*

15

STEAM TRAIN OPERATION

Operating as a common carrier, Hetch Hetchy Railroad was under the rules of the California Railroad Commission. It was required to publish tariffs and timetables, issue passes, submit reports and operate according to standard practices, though the latter was very often ignored by the management and overlooked by state inspectors. The management, as we have said, was unusual, being not railroaders but rather engineering people. In fact, the 1920 report to the state commission shows Mayor James Rolph, Jr., as President of the railroad, O'Shaughnessy as Vice-President and General Manager.

It was a very mixed lot of enginemen and trainmen that operated this standard railroad with its seven miles of sidings and yard tracks, six to eight steam locomotives, various track trucks and buses, and assorted rolling stock. Some were boomers working for a month or two here and there; others were former log train jockeys; several were local men trying to learn the game. Their main job was to get bulk cement up the mountain to the dam construction site, no matter what else happened. And the responsibility for seeing that this was accomplished fell upon Trainmaster Lester B. Cheminant.

Here was a hard-driving man who had no patience with inefficiency, but he was saddled with troublesome equipment and some men who wouldn't have been on the Hetchy if they'd been good enough to hire out to permanent railroads. Chemmie and his family thought of nothing else but cement, according to Mrs. Edith Cheminant many years later. There were even two phones on the table at mealtimes and Chemmie worried constantly. Edith tried repeatedly to keep her husband from being so particular, but they all knew how vital it was to keep up the flow of cement. On some days as much as 2000 yards of concrete were poured at the dam, which meant over 2000 barrels of cement, or 400 tons. Ten carloads would do it, but they had to be there on time and there was very little storage space. On top of this, Cheminant and his family had no way of knowing how long they'd be at Groveland. They had been sent for six weeks, extended to six months. A bond issue would be passed and it would be six months more. As it turned out, they were at Groveland for six years.

There it is: extremely rough mountain conditions, not always the best of men and equipment, inexperienced management, extreme pressure from voters and newspapers. Fortunately, the politicians had to keep their hands out of it or risk the wrath of O'Shaughnessy (they chanced it now and then). The City's newspapers had a field day for years with their typical irresponsible, self-seeking attacks, although some did support the Chief. It seems a little miraculous that such a railroad, owned and operated by a mere municipality, could have accomplished the task. But it did – and it really didn't do too bad a job at that.

Timetables show the speed limits, which varied from time to time: geared engines, 12 mph; rod engines, 20; freight trucks, 18; passenger buses, 20. Tariffs show charges of 12½ cents per ton-mile for carload freight, 17½ cents for less than carload. Passenger fares were 7½ cents per mile. All these details, again, were frills; the only thing that really mattered was getting cement to Damsite and that was what the road's operation was centered on.

Engine No. 3, with a train consisting mostly of cut lumber from Peach Growers, storms across the Tuolumne River bridge a mile below Jacksonville in 1919. Bound for Hetch Hetchy Junction, she has a climb ahead of her to get out of the canyon and over a couple of low ridges. *(—City of S.F.)*

Heisler No. 1 shows off her pilot plow designed by Roadmaster Harwood. Such snow devices as this were effective only when there was a light coating of the white stuff over the rails. Heavy snows got pushed out and compacted to the sides until the engine could push no more and the snow had to be shoveled or otherwise lifted away from the line. The railroad's rotary plow didn't come into use until many years later. *(—Louise Harwood Waldron)*

No. 4 begins the climb up Priest Hill with what was considered a good load: three revenue cars and the caboose. The view is looking up from Moccasin Creek, west of Moccasin Camp. The 3-spot is shown to right, ready to leave Mather, westbound, in 1922. Standing with the small boy are Mr. and Mrs. Pickett and engineer Harley Kennedy, while fireman O'Keefe sits with feet dangling out the cab window. Below is Smith Station after a heavy snowfall in 1922. *(—Above, Earl Hope; right, Mary Male; bottom, Roy Brooks)*

On the curved fill at Rattlesnake Gulch, engine No. 2 pauses for water with her westbound freight train in 1921. Up the canyon to the right is the site of Priest Dam and Reservoir. *(–Bud Meyer)*

While they tried to get ten carloads to the mixers each day, three were about the limit on the Priest Hill grade with one of the big rod engines. Shay No. 6 could do a trifle better and usually struggled along with four loads (one old-timer claimed six loads for her), with maybe an added car of miscellaneous lighter freight or machinery. The railroad conquered this massive obstacle and was able to keep cement supplies adequate to avoid shutdowns over 90% of the time. But again, some highly unorthodox railroading was resorted to.

During the busiest years the lower end of the Hetch Hetchy must have resembled a 27-mile-long railroad transfer yard, stretching from the junction to Groveland. Some of the old trainmen told how it worked. Engine No. 5, the Prairie, would arrive at Hetch Hetchy Junction with the passenger train at 8:30 in the morning, in order to meet the Sierra Railway's westbound train. Having over six hours to kill before starting back, the 5-spot would keep busy hauling loaded freight cars up the grade and out about five miles to Pedro Siding. Eventually a six-car freight train would climb out of the junction with one of the Mikes, No. 3 or 4, pulling and No. 5 pushing as far as the top. Five or six cars which had been spotted at Pedro would be picked up ahead of the Mike and this heavy, twelve-car train with an engine in the middle went very gingerly down a looping two- or three-mile descent into Tuolumne River Canyon. At Six Bit, before the old wooden trestle was replaced, the train would be cut and the Mike pushed the lead six cars across first, returning for the others. Then it was a gentle five-mile climb along the river, past Jacksonville, to Munn Tank (milepost 12) at the bottom of Priest Hill.

Usually eight loads were dropped at Munn and the engine tackled Priest Hill with four. Dud Snider said they'd have to cheat a little if Shorty Morse was running the 2-8-2. He thought four cars of cement were too much of a load and he'd stall every time unless assured he had three of cement and one of, say, hay! The eight cars left behind would be picked up in a couple of six-hour trips of the "shuttle run" – Shay No. 6 from Groveland occasionally got in two round trips in one shift. Having reached Groveland yard, cars were assembled into trains and sent the remaining forty miles to Damsite behind one of the Mikados. Most trains stopped for changes – crews, equipment, consists – at Groveland.

Crews on the upper end would usually stop for chow at South Fork and Mather, though sometimes Damsite could be reached. Freight trains were supposed to be near a place to stop and eat at the end of each four hours. Sometimes it was Buck Meadows Lodge, and often the cars were left in the side track there while the crew

After a three-mile climb out of Mather, eastbound Train No. 4, engine No. 4, passes through the rock cut at Poopenaut Pass and begins the steep descent to Damsite, six miles ahead. *(—Mrs. Emilie McAfee)*

Extra freight train with engine No. 5 passes bus No. 21 in the siding at Damsite, while people pose and tourists take pictures of each other. Hetch Hetchy Valley is in background.

No. 4 gets her eastbound train rolling on the long straight stretch at Buck Meadows, headed for Hetch Hetchy with carloads of bulk cement, merchandise, pipe for tunnel ventilation, and an empty flatcar or two for the Peach Growers sawmill. *(—Mrs. Gladys Little)*

HETCH HETCHY RAILROAD

(City and County of San Francisco, Owner)

TRIP PASS NO. 1564

Pass Mr Calvin Jones

ADDRESS Groveland

ACCOUNT Brakeman HHRR.

GOING TRIP	RETURN TRIP
DATE ISSUED Mar 26 1924	EXPIRES June 30 1924
FROM Groveland	FROM Junction
TO Junction	TO Groveland

VALID WHEN COUNTERSIGNED BY

N. A. ECKART

COUNTERSIGNED BY

N A Eckart
per JH Harwood

M.M. O'Shaughnessy

GENERAL MANAGER

Form 199

HETCH HETCHY RAILROAD

TIME TABLE No. 9

108 19189

EASTWARD—Read Down
In effect May 1, 1922, at 12:01 A. M.

WESTWARD—Read Up
Subject to change without notice

4	6	2	Miles	STATIONS	Elev.	1	5	3
		2:45 p. m.	0	Lv. Hetch Hetchy Junction Ar.	936	8:30 a. m.		
		f 3:08 p. m.	7	Sixbit Gulch Lv.	600	f 8:08 a. m.		
		f 3:25 p. m.	12	Munn Tank	672	f 7:52 a. m.		
		f 3:45 p. m.	17	Cavagnaro	1458	f 7:25 a. m.		
		f 4:10 p. m.	23	Priest Portal	2470	f 7:04 a. m.		
		s 4:40 p. m.	27	Ar. Groveland Lv.	2777	6:45 a. m.		
8:30 a. m.	7:30 a. m.	5:00 p. m.	27	Lv. Groveland Ar.	2777	s 6:30 a. m.	s 4:00 p. m.	s 4:35 p. m.
8:50 a. m.	s 7:52 a. m.	f 5:22 p. m.	33	Big Creek Lv.	2740	f 6:08 a. m.	s 3:38 p. m.	4:10 p. m.
9:00 a. m.	s. 8:00 a. m.	f 5:30 p. m.	35	Smith	3018	f 6:00 a. m.	s 3:30 p. m.	4:02 p. m.
9:20 a. m.	s 8:15 a. m.	s 5:45 p. m.	39	Buck Meadows (b)	3060	s 5:45 a. m.	s 3:10 p. m.	3:45 p. m.
.	f 8:24 a. m.	f 5:55 p. m.	42	Colfax Gate	3120	f 5:35 a. m.	f 3:05 p. m.	
s 9:50 a. m.	s 8:33 a. m.	s 6:03 p. m.	44	South Fork	2875	s 5:25 a. m.	s 2:55 p. m.	s 3:25 p. m.
	f 8:54 a. m.		50	Jones	3810		f 2:34 p. m.	
10:30 a. m.	s 8:58 a. m.	f 6:28 p. m.	51	Intake	3897	f 5:00 a. m.	s 2:30 p. m.	2:50 p. m.
s11:00 a. m.	s 9:25 a. m.	s 6:55 p. m.	59	Ar. Mather Lv.	4520	4:35 a. m.	2:00 p. m.	2:15 p. m.
7:30 a. m.	June 1 to		0	Lv. Yosemite Valley (Y.T.S.) Ar.		June 1 to	5:45 p. m.	
11:00 a. m.	Oct. 1		33	Ar. Mather Lv.		Oct. 1	2:00 p. m.	
11:15 a. m.	9:30 a. m.	7:00 p. m.	59	Lv. Mather	4520	s 4:30 a. m.	s 1:50 p. m.	s 2:00 p. m.
	9:38 a. m.		62	Summit	5064		1:40 p. m.	
s12:00 m.	s10:00 a. m.	s 7:30 p. m.	68	Ar. Damsite (Hetch Hetchy) Lv.	3869	4:00 a. m.	1:15 p. m.	1:20 p. m.

(b) Passengers for Bower Cave may secure transportation to or from Buck Meadows through arrangement with Hetch Hetchy Railroad Station Agents.

ABBREVIATIONS: s Regular Stop. f Trains stop on signal only. Ar. Arrive. Lv. Leave.

Owing to limited space, the Railroad does not guarantee immediate transportation of baggage on motor cars.

Connections at Mather to Yosemite Transportation System's "Tioga Pass Route" between Yosemite and Tahoe.

Trains No. 1 and 2 connect with Sierra Railway, Southern Pacific and Santa Fe Railways.

Trains No. 1, 2, 5 and 6 arrive at and depart from passenger depot at Hospital Road Crossing, Groveland.

Trains No. 3 and 4 arrive at and depart from Freight Shed.

L. T. McAFEE, SUPERINTENDENT

L. B. CHEMINANT, ASST. SUPT. & TRAINMASTER

1924 No. 3

Hetch Hetchy Railroad

Pass L. T. McAfee

Account Supt., H. H. R. R.

Between all Stations

UNTIL DEC. 31, 1924 | UNLESS OTHERWISE ORDERED AND SUBJECT TO CONDITIONS ON BACK

M.M. O'Shaughnessy

GENERAL MANAGER

Form No. 2

Form H. H. W. S. No. 62

Locomotive { Number..........
Initials..........

HETCH HETCHY RAILROAD

Locomotive Inspection Report

Instructions.—Each locomotive and tender must be inspected after each trip or day's work and report made on this form, whether needing repairs or not. Proper explanation must be made hereon for failure to repair any defects reported, and the form approved by foreman, before the locomotive is returned to service.

Inspected at.........................., time..........m. Date..........191

Repairs needed:

Condition of injectors.......................... Water glass..........

Condition of gauge cocks.......................... Brakes..........

Condition of piston rod and valve stem packing..........

Safety valve lifts at..........pounds. Seats at..........pounds.

Main reservoir pressure..........pounds. Brake pipe pressure..........pounds.

(Signature)..........

(Occupation)..........

The above work has been performed, except as noted, and the report is approved.

.........................,
Foreman.

HETCH HETCHY RAILROAD

CONDUCTOR'S TICKET

Form H. H. W. S. No. 48 | 3891

Date		Station	From	To
		H. H. JUNCT.		
		WEST BRANCH		
1919		SIXBIT GULCH		
1920		MUNN		
1921		STEVENS BAR		
Jan.	July	CAVAGNARO		
Feb.	Aug.	RATTLESNAKE CREEK		
Mar.	Sept.	PRIEST PORTAL		
Apr.	Oct.	BIG OAK FLAT		
May	Nov.	GROVELAND		
June	Dec.	BIG CREEK		
1	17	SMITH		
2	18	RANGER		
3	19	BUCK MEADOWS		
4	20	SOUTH FORK		
5	21	RED HILL		
6	22	JONES		
7	23	INTAKE		
8	24	HOG RANCH		
9	25	CANYON RANCH		
10	26	DAMSITE		
11	27			
12	28			
13	29			
14	30			
15	31			
16	*			

Stations FROM and TO to be designated by Conductor's punch marks.

PUNCH HERE 1/2 IF FOR HALF

M. M. O'Shaughnessy
GENERAL MANAGER

On the railroad's forms the initials H.H.W.S. stand for Hetch Hetchy Water Supply. The Conductor's Ticket at right is from the collection of Vernon Sappers.

The views above show a typical train at Mather in winter season: wedge plow No. 101, engine No. 4, boxcar and caboose. After arriving from Groveland, the engine would customarily push the plow along various sidings and spurs in order to keep snow from piling in drifts. *(—Dud Snider)*

rode their engine up near the lodge, leaving it on the main line while they ate! An interesting bulletin of 1919 reads: "Train Conductors and engineers: Mose Baker, Jr., Jack Male, F. H. Parker, Thomas Duran, A. R. Sloan, H. Kennedy, F. L. Hotchkiss, R. J. Irvin. Upon taking oil at any point where you are to eat dinner, or where you are delayed for any other cause, you will always attempt to have the oil running in the locomotive tank while you are eating . . ." being careful to avoid overflow. It can be imagined that up in those altitudes in wintertime the heavy fuel oil flowed very slowly indeed. Track bus No. 19 was usually dispatched with driver Frank Dennis to pick up and transfer crews when the 16-hour law caught up. Peculiar railroading, but it was flexible.

Of the locomotive engineers named, Al Sloan came over from the Peach Growers rail operation to run the Hetchy's Heislers while logging operations were shut down in winter. Harley Kennedy had worked as a hoghead on Westside Lumber's narrow-gauge Hetch Hetchy & Yosemite Valley Railroad out of Tuolumne City and developed a reputation as the finest engineman on Hetch Hetchy, "a real master with the air on hills." Tom Duran, on the other hand, didn't hold his train so tightly and opposing track buses allowed themselves plenty of time to get in the clear when they had a meet with Duran on a descending grade. Engineer Hotchkiss later transferred from geared engines to bus No. 23 on the dam run. Others who ran locomotives later on the Hetchy were Shorty Morris and Tom McKenna; firing for them were Neil Sinclair, Mickey O'Keefe and several of the local lads. Herman Nagel was an experienced brakeman, spending two or three years on the passenger run with Mose Baker. Herm refused to try for promotion to conductor because that meant he'd have to go on night runs on "them old Shays" hauling cement up the hill.

Train dispatching was done from an office at Groveland. Men holding down the chair included Earl Durgan, J. E. Smith and a man named Pickett; extra or relief men were Ray Carne and Cal Jones. Train crewmen who had families usually set up housekeeping at and near Groveland; single men took rooms. The engineers had various individual ways of sounding the whistles when arriving with a train at the home base to let their wives know they'd be expecting a hot meal in a half hour or so. Speaking of whistles at Groveland reminded Ernie Beck of the night he was at work in the shop-roundhouse and spotted a glow in the sky up towards town, perhaps half a mile to the west. He jumped into the 4-spot's cab and sounded four long blasts on her whistle, then four more. Lights came on all over town and the volunteers ran to fight the fire at the Europa Hotel. The hotel burned to the ground, but a great catastrophe was avoided.

When the passenger train was steam operated, the 5-spot left Damsite with Train No. 1 at 4:00 a.m. daily and covered the 68 miles to Hetch

Shay No. 6, with a cement train on Priest Hill, pauses for a brief rest at the Rattlesnake Gulch curve. After one dragged-out trip in 1923, engineer Hotchkiss made the following notation on the back of his Engineer's Trip Report, form 38: "Delayed acct. of Eng. 6 coming to pieces." *(–Earl Hope)*

Hetchy Junction in four and one-half hours, reaching there at 8:30 for connection with the Sierra train which took passengers on to the SP or Santa Fe at Oakdale. After the Sierra's return, Hetchy's Train No. 2 left the junction at 2:45 p.m. and tied up at Damsite at 7:30. It was a long day for the crew, but this was improved in wintertime, when Trains Nos. 1 and 2 ran only between Groveland and the junction, Nos. 3 and 4 covering the upper end of the line. While this schedule was in effect, passengers stayed overnight in Groveland.

Such arrangements were certainly not designed to attract passengers, and there developed over the years fierce competition from autobuses, competition that ran all the way from Stockton and bypassed the SP branch train, the Sierra Railway and the Hetch Hetchy, while making far better time – and avoiding all those changes! It's not surprising that Hetch Hetchy passenger service was soon converted to track buses only; these were adequate for the number of people to be carried.

Early in 1920 the Hetch Hetchy Railroad, in conjunction with Sierra Railway, was trying to attract a substantial Yosemite passenger business over the route. However, they had to be able to get people to Hetch Hetchy by the evening of the same day on which they left the City, then send them over to Yosemite the next day. The Southern Pacific was asked to speed up its running time between San Francisco and Stockton and down the branch to Oakdale (there was a through coach), since it took over six hours for this 124 miles alone. SP was able to cut over an hour off the time, and the Hetch Hetchy could get people to the dam by 7:30, only 11½ hours after leaving the City! But the hordes of passengers failed to materialize, even though the scenery "compares favorably with the Canadian Pacific and Royal Gorge railroads," as stated in the 1920 annual report.

Starting in September 1921, the railroad had a contract to carry U.S. mail, serving post offices at Big Oak Flat, Groveland, Buck Meadows, Mather and Hetch Hetchy. Oakland Recreation Camp on the Middle Fork was added in summer months.

Winter snowstorms slowed everything on the Hetch Hetchy Project, even the cement trains at times. For many months the entire line beyond Moccasin was subject to icy rails, especially in gulches and canyons. Extra sand was carried to help maintain traction and in braking. The first true snowplow was built, using a loaded gravel car on the front of which was fastened a heavy corrugated-iron wedge to shove snow aside. A cabin was provided on top for the "lookout." This rig and a later more orthodox wedge plow worked fine in light snow, but they were useless in a big storm where snowdrifts needed to be lifted and thrown away from the line. (The renowned and unique rotary snowplow wasn't put into service until many years after the initial construction phase.) Sometimes the tracks would be closed for days and every available man was out, all working like beavers to get trains running again. Cement, yes, but food as well and other supplies had to be taken in to the upper construction camps where there was no road in winter but the railroad.

In the annual report for the Bureau of Engineering for 1921-22 there is a section that gives the picture. During the seven days before Christmas ten feet of snow fell at Summit "and only the use of a new snow plow, which had been lately constructed in the Groveland shops, prevented actual blocking of the railroad." And there were heavy slides in the vicinity of South Fork. By the following year, however, the railroad was able to report that improved roadbed and equipment made it much easier to keep traffic moving.

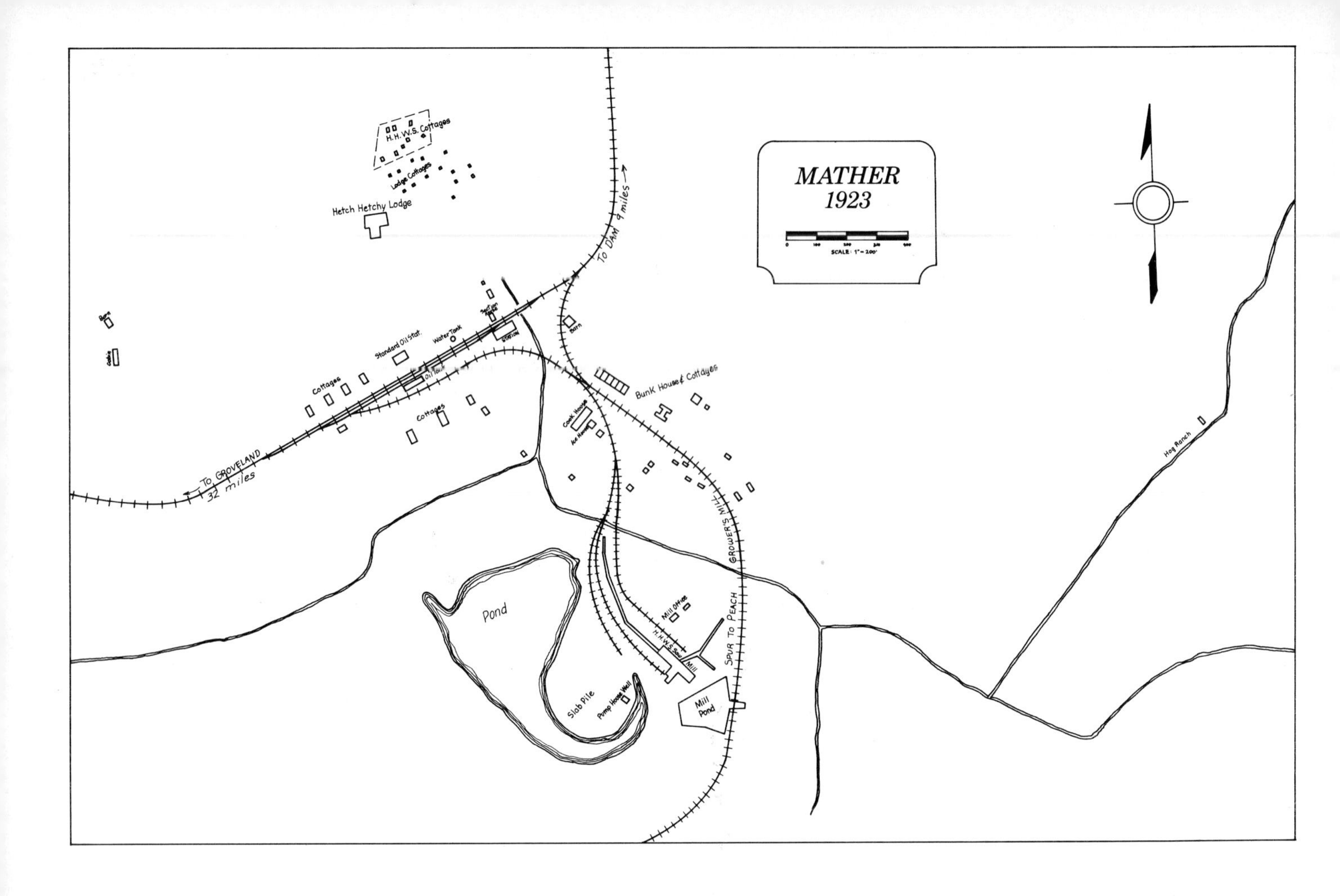

Track layout at Mather shows Hetch Hetchy Railroad to Damsite at top and lumber railway of California Peach Growers heading south at the bottom. The wye near the station was added in later years when it became necessary to turn locomotives at this point. The mill pond is now a swimming lake for Camp Mather summer resort.

The Groveland map below is a composite of different periods centered on 1920. The wye to right was constructed when the turntable proved inadequate for heavier locomotives. A large part of this area, from Big Oak Flat Road across to the hospital site, was selected in 1973 to become Groveland's town park.

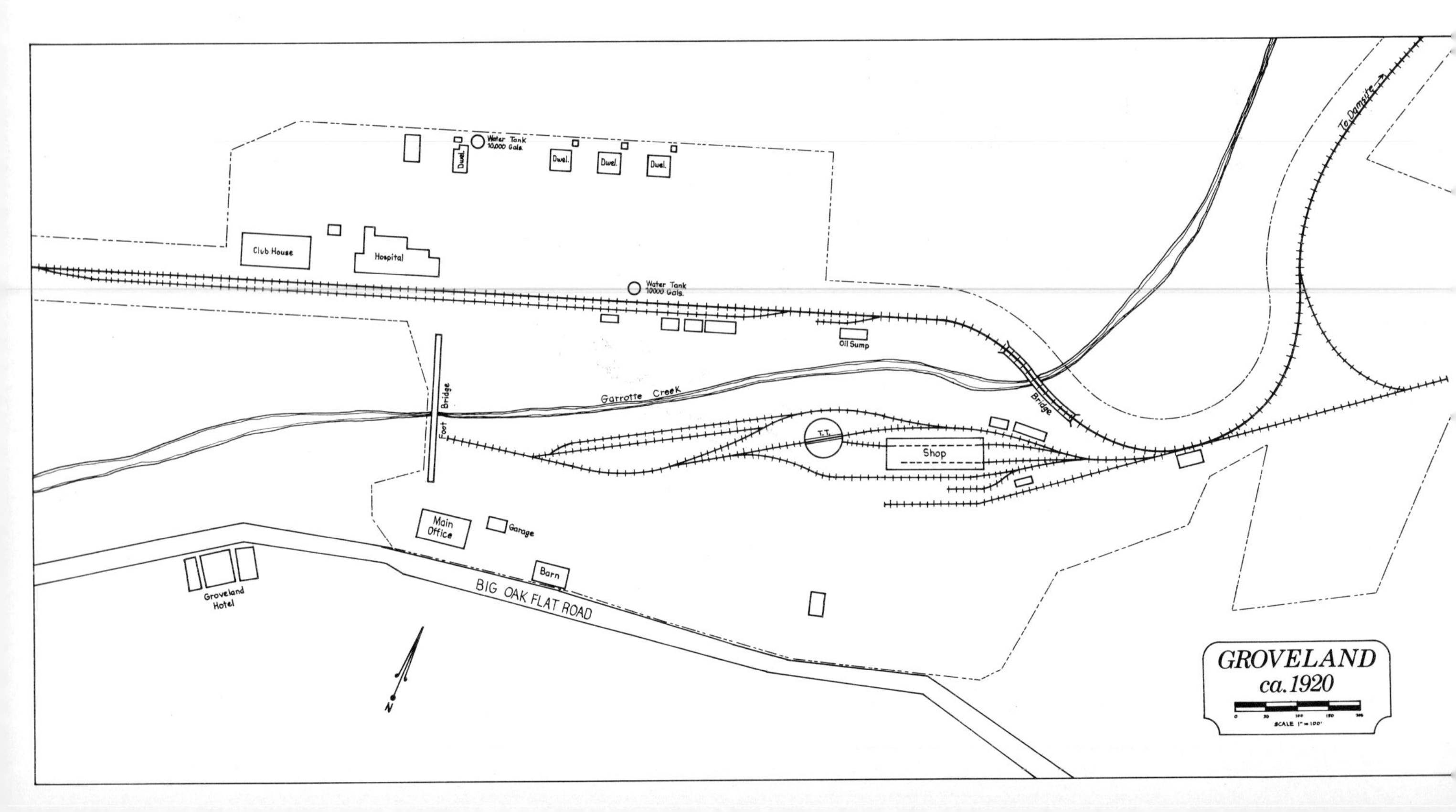

16

EXCURSIONS, WRECKS AND WORMS

Tour buses in the early 1920s left Yosemite Valley at 8:00 each morning via Big Oak Flat Road and reached Mather about noon. Sight-seers transferred to track buses for a grand scenic ride to Hetch Hetchy Valley, where they were given about an hour to look in wonderment at various phases of the work in progress, then returned to Mather. Here the park authorities had opened Hetch Hetchy Lodge and cabins so that some of the tourists were able to stay over a day or two, if they wished, while the rest returned to Yosemite. In 1919 and 1920, at least, the Yosemite Park Company furnished and operated the track buses themselves and paid San Francisco $1.50 per passenger for use of the track. In 1921 there were 60 people a day making the trip, and a year later, 80.

This Hetch Hetchy tour got so popular that in 1923 and 1924 the railroad operated a steam train on the run, using open-side excursion cars Nos. 6 and 7. The 1924 service, which continued up to September 6, constituted about the only railroad operation east of Groveland that year. The last summer for the Yosemite excursions was 1925, with 1748 passengers being carried in track buses. Two years later the park's Hetch Hetchy Lodge would be bought by San Francisco and opened as part of a new summer resort.

Other special trains and excursions ran on the Hetch Hetchy, some quite unusual in keeping with the tradition. Oakland and Berkeley old-timers fondly remember riding the trains when going to their summer camps on the Tuolumne River. Oakland campers elected in 1922 to use train service and avoid an overnight stop at Knight's Ferry, as had been required with the previous year's bus. Round-trip rail fare was $11.25, the train leaving Oakland each Saturday morning with as many as 125 passengers. Berkeley people used the same train in 1923, the year after their camp opened at Hardin Flat. A steam train carried the groups as far as Groveland, where they transferred to track buses for the final 17 miles to South Fork. From there it was just a four-mile ride up Big Oak Flat Road to Camp Berkeley, whereas Oakland Camp was just around the bend from South Fork and alongside the tracks. Ray Carne recalled driving bus No. 21 loaded with the Oakland gang, "at times so many passengers I could hardly handle the controls.... The next trip we might have a load for San Francisco Camp at Mather, and there pick up the Yosemite tourists for the run to Damsite and back, then return to Groveland as Train 3."

Biweekly excursion trains were operated from Oakland Camp to Hetch Hetchy to view heavy construction during 1921-23. This was a very popular two-dollar trip. The last year in which the railroad served the summer camps was 1924, since work at the upper end had been completed and there were no scheduled trains east of Groveland. But a special through service was operated during the 1924 season between the junction and Oakland Camp on Saturdays only.

The City's railroad throughout its busy early life ran all sorts of interesting excursion trains for politicians, businessmen, even the taxpaying general public. One of the great ones was the July 1920 trip of the Electrical Development League, which left San Francisco Friday evening in special Pullman cars on Santa Fe train No. 2. The cars were detached at Riverbank, taken to Oak-

Derailments were common enough on the twisting Hetch Hetchy Railroad. Here, everything has been put back in place with the help of one of the Pierce-Arrow track trucks, on the grade near Big Oak Flat.

Mikado No. 3 with train of lumber, westbound, gets ready to stop for a drink of water at the Rattlesnake Gulch tank. In background, work has begun on Priest Dam. *(–City of S.F.)*

"Grand Jury Special," engine No. 5 with caboose and coach 11, waits at Damsite on May 30, 1921, while Grand Jurymen from the City inspect progress at the construction site in the canyon beyond. *(–City of S.F.)*

Special train to dam dedication ceremony at Hetch Hetchy on July 7, 1923. Engine No. 5 waits with her train at Mather. Combination car No. 11 had been purchased from the Ocean Shore line out of San Francisco; the other cars were open observation cars Nos. 6 and 7. *(—Vernon Sappers)*

Trains carrying sight-seers from Oakland and Berkeley summer camps made the run from South Fork to Damsite twice a week. No. 5, above, at Mather in 1921, draws a caboose, roofed flatcar with benches and regular coach. To left, on the way to Damsite in 1922, a Heisler locomotive has a rare two-caboose train with loaded gondola between. *(—Above, Mr. & Mrs. R. O. Richardson; left, Oakland Recreation Dept.)*

dale and passed on to the Sierra Railway, arriving at Hetch Hetchy Junction early Saturday morning. Riding in the Hetchy's "special observation car," the party stopped to inspect various construction sites along the line, reaching Damsite and the valley at 4:15. The return trip left Mather Sunday morning at 8:00, rode the scary tramway down to Early Intake and back, and inspected South Fork Portal and all the sights of Groveland – which took all day! At last the special departed Groveland at 7:00 Sunday evening and reached the City 12½ hours later via Santa Fe Train No. 11 and the ferry *San Pablo*.

There were special trains carrying inspection parties of San Francisco Supervisors and the Grand Jury – "parties" is the appropriate word to describe these grand junkets. Even during Prohibition there was plenty of booze, though sometimes it had to be hidden aboard the train and sneaked past park rangers just outside Mather. Regular week-end excursions were operated in 1923 for all sorts of groups from the City, to acquaint people with the immensity of the project and show where their money was being spent. A fare of $29.50 paid for the entire trip: Pullman berth, meals, everything. Saturday night was spent at the Park Company's Hetch Hetchy Lodge and on Sunday evening the groups returned to their Pullmans at the junction, arriving in San Francisco early Monday. The guide on these tours was former railroad manager Cheminant. He was ideally suited for the role, according to many who went along, for he had a tremendous knowledge of the project and an intense love of the mountain country. Chemmie also had a way with words; he authored the engineering department's thick and highly readable annual report books.

Mr. Charles Clerk, who retired many years ago as conductor on the famous *California Zephyr*, was employed as a Pullman conductor in the '20s and happened to get assigned to two special trains to Hetch Hetchy in 1923. The first trip was all politicians, with Chief O'Shaughnessy, but the second was a great special to the dam dedication of July 7, 1923. Pullmans were left at the junction, but Clerk was invited along for the ride to "Hooch Hoochy" and had a great two-day trip with the gang, including drinks in the baggage car out of Mayor Rolph's personal flask. After ceremonies at the dam, the entire party stayed overnight at the lodge, were given a thrilling ride next day down the Intake tramway and eventually reached home sometime Monday. It was a trip that stayed vivid in Charlie Clerk's mind as long as he lived.

The Hetch Hetchy had a remarkably clean record as far as fires and all forms of pollution were concerned, but like all steep and crooked rail lines there were numerous derailments, a few minor runaways and a couple of unfortunate head-on collisions. However, the spectacular runaway of September 13, 1922, was enough for the road's entire life span.

Engine No. 4 came out of the junction with five loads of cement, then coupled on behind four more loads at Pedro Siding and started downgrade toward the Six Bit trestle. A new brakeman, Tom Fleming, was working that day and it is assumed that he failed to connect the air to the cars ahead. It didn't take engineer Tom McKenna long to discover that he couldn't hold the train – it felt as though the 4-spot were being stretched out between the rolling tons ahead and the dragging braked cars behind. McKenna blew a series of blasts on the chime whistle, alerting his conductor to jump and warning a crew working at the trestle far below to get clear. Then McKenna and fireman Tom Duran leaped clear, sustaining only slight injuries. Fleming could be seen trying to set hand brakes as the train hurtled downgrade and crashed off the tracks in a cut between Red Mountain Bar and Six Bit.

The wreckage caught fire, and poor Fleming was moved to safety but died shortly afterward. Superintendent McAfee gave orders to let the fire burn out – he didn't want any water poured on those tons of loose cement blocking the tracks. No. 4, bruised and burned, was pulled up to the shop, where her rehabilitation took a couple of months, during which time the Northwestern Pacific's Shay was rented. Later No. 4 was involved in an unusual derailment above Priest. They had to rig up a deadman to keep her from rolling all the way over, and she stayed like that for three days while Ray White was detailed to remain with her and keep the fire hot.

Serious accidents were rare on the Hetch Hetchy Railroad. In the one shown at top, Midland Valley boxcar No. 3542 was soon righted and the roadbed between Groveland and Big Oak repaired in a few hours. The two lower views, however, show what happens when a loaded train runs away on a steep grade. Engine No. 4, with cement cars ahead and behind, started downgrade at Red Mountain Bar. Apparently the air brakes hadn't been connected to the cars in front and they dragged the train down to a crashing mess just short of Six Bit Gulch bridge. In center, No. 4's tender is pulled back by engine No. 5; the wrecked engine has been shored up with ties to keep her from falling over. Below, a view of the jumble made when four boxcars laden with bulk cement are derailed at high speed in a cut.

When No. 4's cement train came to a sudden halt near Six Bit, fire broke out in the wreckage; boxcars and engine were badly scorched. In the view above, men are shoveling the salvaged cement into sacks, thankful that the accident had taken place in September and not in the rainy season. In the background, cars from the rear of the train are pulled back toward Pedro Siding by engine No. 5, which had been at Hetch Hetchy Junction for the afternoon passenger run. No. 5 is shown in lower photo at work on wreckage clearance within a few hours after the accident. *(—Both, City of S.F.)*

Hetch Hetchy shopmen at Groveland repaired No. 4 after the accident. Starting at right corner: Ed Coyle (back to camera), Walt Magee, Bert Minard and Frank Schott (with hat); others not identified. *(–City of S.F.)*

There was a cornfield meet at Big Creek when boxy express bus No. 22 was practically demolished by Mikado No. 3 with a freight train. A couple of passengers were injured when another track bus ran into a train just west of Mather. A speeder operator was killed running head-on into bus No. 19 on the sharp curve just east of Buck Meadows. The Lord knows how many such accidents were avoided by quick stops and rapid retreats to the nearest siding; we've been told of a couple of close ones with full trains involved. The attitude of most rails was not to report it if no damage was done, and that's the way it worked on the Hetch Hetchy.

It's a cinch management wasn't consulted the time Tom Duran gave help to a Sierra doubleheader with his engine No. 6. The Sierra train was unable to start on the upgrade at Hetch Hetchy Junction and Duran talked them into letting him have a try with the big Shay. Butch De Ferrari said the Shay took right off with the train, no trouble at all; he didn't say how it got back – maybe it was pushing from the rear. Several others have mentioned assisting the Sierra with a shove from the rear at the junction, unofficially of course.

Still, even old man Cheminant was known to bend the rules from time to time, as long as it didn't interfere with his cement trains. One Christmas Eve he was aboard a train leaving Damsite and the crew made sure they departed on schedule. However, about a thousand feet out from the station there was a mysterious 45-minute delay caused by a hotbox on one of the car trucks. This happened to be just long enough for the working shift to get off duty and catch the train to town. On Sundays there appeared to be considerable leniency and almost any of the railroad em-

Locomotives Nos. 4 and 6 stand outside the shop at Groveland, while No. 5 peeks out the door behind the big Shay. This was early on a summer morning in 1922; the photo was taken by Ernie Beck, night hostler and call boy, as he was going off duty after getting the engines ready and calling the crews.

ployees could get a sort of "Sunday pass" to take themselves and visitors up to see the big show at Damsite or down to Stevens Bar for fishing or a swim. One former employee said he never had any trouble borrowing a speeder on Sundays for personal fishing trips up to South or Middle Fork, or even to Damsite, where there was wonderful fishing upstream from the construction.

Hetch Hetchy Railroad made news in the Oakland *Tribune* one day in 1922, when a headline proclaimed "Jet of Steam Puts Ambling Worm on the Run." The story told of an army of queer-looking "caterpillars" which covered the rails for a mile and stalled all trains. Sand had no effect, so scalding steam was sprayed ahead of the locomotives to drive the invaders away and degrease the rails. On another occasion a train was descending Red Hill and started sliding on rails thoroughly greased by squashed insects. Fortunately, it was a slow slide to the bottom and no great harm was done, but five cars had to be set out at South Fork for repairs. Men with brooms were stationed on the front of the next train so the oily residue would be lessened.

Railroading on the Hetchy was always a rugged job. It was terribly hot in summer months, particularly on lower stretches of the railroad and climbing the sun-blasted side of Priest Hill. Shay No. 6 got so hot inside her cab on this climb that both engineer and fireman would often get out and ride on the tender. Then, as a shocking contrast, Cappy Caplinger told of a problem with the gasoline rigs in winter, when the temperature often went below zero: he had water freeze in radiators, even when running! They'd always throw an old blanket or something over the radiator when starting out and often even removed the fan belt to let engines run warmer.

Les Phelan told of the time he was taking the boss, McAfee, east on a speeder in a snowstorm. Reaching Colfax Springs, they could get no farther, nor could they retreat back up the grade. Mac jumped off and chanted, "The Snow, the Snow, the Beautiful Snow – up to your ass wherever you go." Then he said, "See you later," and started back on foot to the nearest camp, possibly Big Creek, eight miles.

During fiscal year 1924-25 the system's railroad traffic had decreased rapidly as successive units of construction were completed from Early Intake to Moccasin. Steam passenger train service between Mather and Damsite was discontinued on September 6, 1924, and all passenger service east of Groveland was stopped effective January 1, 1925. Six weeks later, on February 15, the Hetch Hetchy Railroad ceased operations as a common carrier, and all scheduled service was discontinued on March 21, on which date the postal department released the City from its mail contract. Between September 1 and November 10, 1925, track was removed from Damsite back to Mather, using engine No. 5, and the former rail line was surfaced as a roadway, as specified in the Raker Act.

The management of the California Peach & Fig Growers lumber outfit had realized that the Hetch Hetchy Railroad was to have a limited life

Under terms of the Raker Act, HHRR tracks had to be removed between Mather and Hetch Hetchy Valley when dam construction was completed, and the right of way converted into an auto roadway. Here is No. 5 with the track-removal train in 1925. *(—Roy D. Graves)*

span as a common carrier. But when the City requested permission to end this service, there was a large backlog of cut timber to be sawed at the Mather mill and some lumber to be shipped. A contract was signed by the Peach Growers and the City to provide for removal of three million feet of lumber over Hetch Hetchy rails before July 31, 1925. The lumber operators, who were responsible for all track maintenance east of Groveland, rented Hetchy engine No. 5 and hired a train crew, including Herm Nagel as conductor, to work for three months getting this material out. Starting from Groveland, they'd go up to Mather one day, haul six loads to Hetch Hetchy Junction the next and tie up again back in Groveland. Del Gilliam said they were still taking out some lumber for the same people in 1927, because he was hired during high school vacation to run a speeder as fire patrol exactly twenty minutes behind the trains, as required by a special Forest Service rule.

Keeping in mind that the Hetch Hetchy Railroad was never designed or expected to make money, that it existed primarily to make an expensive hauling job less expensive, it is interesting to check the road's reports to the California Railroad Commission. Net operating losses varied from a high of $240,203 in 1922 to a low of $54,181 in 1923. There was actually a small profit in 1924, when work on the upper end neared completion. The busiest year was 1923, during which the railroad carried 145,000 tons of freight and 48,000 passengers. Just three years later, when heavy construction was over and the line was no longer a common carrier, freight haulage had dropped to a mere 6000 tons and passengers to only 313. So much for the earlier grand plans for a permanent Hetch Hetchy Railroad serving as the alternate route to Yosemite and carrying out lumber.

It is appropriate at this point to leave the railroad for a few chapters and catch up with progress on the aqueduct line. We have seen how the big dam was finished and the 19-mile tunnel opened from Intake to Priest, but much more work had to be done before power could be generated and water brought into San Francisco. We'll pick up the story at Priest and return later to the Hetch Hetchy Railroad — for this great line definitely did not die when initial construction was finished. It lived on and worked through the 1930s and the war years to expire finally in 1950.

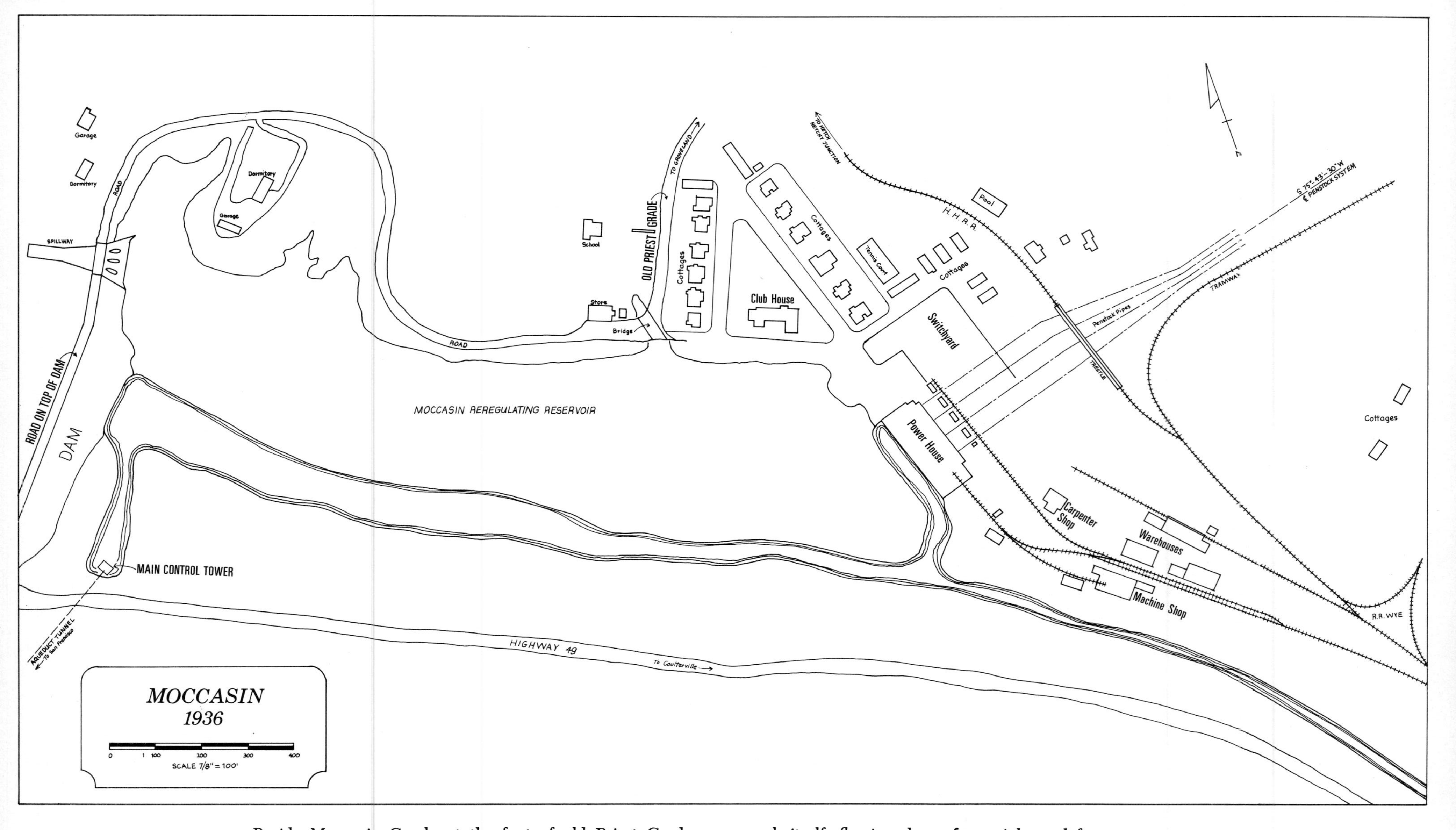

Beside Moccasin Creek, at the foot of old Priest Grade, the City of San Francisco placed its first major powerhouse. Here had been old Hughes stagecoach station near the intersection of Big Oak Flat Road and the road to Coulterville; a covered bridge crossed the Moccasin. The creek itself, flowing down from right to left, now passes beneath Moccasin Reservoir in a conduit. Thus, potentially dirty water cannot get into the City's drinking supply. After 1930 Moccasin became the center of operations for Hetch Hetchy Railroad.

17

PRIEST, MOCCASIN AND DISASTER

The first big tunnel was completed and water passed through from Early Intake to Priest on June 2, 1925. But before that could happen there still remained the task of preparing to convert water power to electricity. The job at Priest was to provide a holding place for the water so that it could be released in varying amounts to a powerhouse at the bottom of Priest Hill. During peak-use hours, demands for electricity are naturally increased and large amounts of water are required to generate this power when needed, because such electricity cannot be stored. The storage reservoir, or forebay, above a powerhouse is drawn down in peak-use hours, then the level rises again the balance of the day as water continues to surge in at a steady rate.

The Mountain Division Tunnel ended at Priest Portal, elevation 2169, near the bottom of Rattlesnake Creek Canyon. Here it was necessary to construct Priest Dam, which would be the fifth major dam of the project, following those at Lake Eleanor, Cherry Creek diversion, Hetch Hetchy and Early Intake. This rather large structure would create a forebay for Moccasin Powerhouse. A power tunnel had to be excavated, extending from the western side of this reservoir; it would carry water just over a mile to the head of two giant steel pipes, the penstock. These would drop steeply down the face of the hill, right into the powerhouse, to build up water pressure and turn four generators with a force capable of developing about 100,000 hp. When the water emerged from the powerhouse, it would eventually renew its gravity flow to the homes of San Francisco.

In the fall of 1921 construction had started on a camp at Priest Portal which housed approximately 300 men who were building Priest Dam and digging out the power tunnel across the canyon. At the same time a similar camp was started below in the valley of Moccasin Creek and a couple of smaller ones along the penstock route between. The first equipment put to work at the Priest Reservoir site was a revolving steam shovel, two 20-ton dinky tank locomotives and a flock of four-yard dump cars of three-foot gauge. With these, excavation began for an approach to the power tunnel and for material to be used as fill for the dam. Equipment was also installed to sluice earth from adjacent hillsides and place it to form the upstream portion of Priest Dam. Then, back up Rattlesnake Creek near the Priest Hotel, there was an old mining-days diversion dam and tunnel to turn drainage under the road and into Grizzly Gulch; the City had to repair and enlarge this facility to keep contaminated water out of the new reservoir.

Priest Dam was completed in 18 months by day-labor employees of Hetch Hetchy. It is 1160 feet long and 148 feet high, consisting of an articulated central concrete core wall to bedrock, earth fill in the upstream zone placed by hydraulic methods, and earth fill downstream deposited by the dinky locomotives and dump cars and sluiced into place with water jets. Spoil from the aqueduct tunnel forms a downstream toe, the face upstream being paved with hand-placed rip-rap to prevent erosion from wave action. Behind this new dam rose a reservoir capable of holding 2350 acre-feet of water, about two days' flow of the big tunnel.

A few of the former employees remembered that when the Priest job neared completion, the

Near Priest, on top of the hill above Moccasin, a dam was constructed to form the forebay reservoir for Moccasin Powerhouse. These 1922 views show work in progress as the large earth dam was formed. Above, looking eastward across Rattlesnake Gulch, the camp in left distance is placed above Priest Portal of the 16-mile tunnel from Early Intake. The camp at upper right houses dam builders, and a three-foot-gauge incline railway descends the hillside from this camp to the dam. The tunnel rail tracks cross center of picture in order to bring supplies of granite for the dam face. Across the top is the line of Hetch Hetchy Railroad. Below, earth fill is being dumped to form the downstream portion of Priest Dam. The concrete core wall takes shape in lower right; upstream from the core, more earth fill was sluiced in from near-by hillsides. *(—City of S.F.)*

Over-all view of Priest Dam construction, looking westward across Rattlesnake Gulch. To left, narrow-gauge dinkies and dump cars bring earth fill for placing and compacting. Next, the concrete core wall is being poured from tiny push cars running on rails along the top of forms. The two tall trestles in center carry pipes through which muddy earth is being sluiced in from the opposite hillsides to form the upstream portion of the dam. The view below, in September 1923, shows the complete dam in place. Its upstream face is being covered with fitted granite rock removed from the aqueduct tunnel. At the top of the dam sit two dinky locomotives and their dump cars, now idle and awaiting the next assignment. *(—City of S.F.)*

One of the three-foot-gauge Porter 0-4-0 tank locomotives with its string of dump cars, as Priest Dam nears completion. Nearer, in center, stonemasons carefully place granite pieces from the tunnel, forming a firm face for the upstream portion of the dam.

An aerial view of Priest Dam and Reservoir, shortly before abandonment of Hetch Hetchy Railroad in the late 1940s. The white structure to left is the outlet tower, controlling the flow of water to the powerhouse at Moccasin. The railroad grade can be seen at bottom, climbing and twisting up the sides of Rattlesnake Gulch and back across the top as it makes its way toward Groveland. *(—Both, City of S.F.)*

Chief left for a vacation in his native Ireland. Strict orders were given that no water would be accumulated in the reservoir until after he returned. Some well-meaning engineers went ahead on a test and impounded the water — and the core wall was bent very slightly downstream by the pressure. My God, there was hell to pay! But additional material was compacted against the downstream face and the dam was watched and measured year after year by state inspectors. It was found to be entirely stable and has survived to do its job for half a century.

Priest Reservoir connects with the Moccasin Powerhouse penstock by means of a mile-long tunnel, 13 x 13 feet and in cross section horseshoe-shaped, with a capacity of 800 m.g.d. Huge steel penstock pipes were extended from a surge chamber at the west end of this tunnel, emerged from underground at West Portal construction camp, and were fastened to the face of Priest Hill for a vertical drop of 1300 feet in about one mile of pipe length. The tunnel portal was completed with connections for three penstock pipes, one of which was capped for future expansion. Two 104-inch

Water flowing out of Priest Reservoir passes through a mile-long tunnel and emerges at the top of the hill above Moccasin. Here is the west portal of this tunnel during construction in 1923.

Early in 1925 penstock pipes were being placed at the same location. Pipes were carried up the hill on a standard-gauge cable incline railway, shown at lower right, and placed on the concrete saddles. *(—Both, City of S.F.)*

pipes started the descent, soon branching into four smaller ones for the main drop, and then near the bottom into eight 36-inch pipes which entered the powerhouse.

Penstock construction was an extremely difficult job because there were 6,200 tons of heavy pressure pipe to be brought in, raised up the hillside and put in place, then held there permanently. The job took a year to complete and involved some very interesting rail operation. Late in 1921 the spur track was put in, leaving the Hetch Hetchy main line three miles up the hill from Munn Tank. This branch line, 7000 feet long on a 5% grade, curved and twisted down to the powerhouse site near Moccasin Creek. During the declining years of the 1930s and '40s Moccasin was destined to be headquarters for the Hetch Hetchy Railroad, but at first only enough tracks were laid in to carry materials for building the powerhouse and the penstock. To operate on this difficult trackage, since all HHRR locomotives were busy, Shay No. 12 was once again leased from the Sierra Railway and stationed at Moccasin.

The purpose of penstock pipes is to carry water under ever-increasing pressure down to the powerhouse, using the tremendous force to generate electricity. Looking down from near West Portal Camp the view shows almost a mile of pipeline dropping 1300 feet to the powerhouse. In foreground, two massive pipes emerge from the tunnel and are divided for the drop. The Coulterville-Sonora Road, later Highway 49, crosses picture at top. *(—City of S.F.)*

Penstock construction at Moccasin, looking uphill. Hetch Hetchy Railroad crosses on the trestle; to right, the incline cable railway in turn crosses the railroad on another trestle. *(—City of S.F.)*

Looking down the penstock line during construction in 1923. A car is being pulled up the cable incline to left. In the very upper right corner Big Oak Flat Road branched off from the Coulterville Road.

San Francisco's Grand Jury, on a tour of inspection in 1924, carefully posed on the Moccasin incline railway car, ready for the ascent along the line of penstock pipe construction. Photographer Chaffee, with his huge 5 x 7 plate camera, stands to left of car. *(—Bill Firmstone)*

While work was in progress on the penstock line, the powerhouse on Moccasin Creek below was built. Here, in September 1923, foundations are being poured for the massive structure. *(–City of S.F.)*

The 12-spot delivered pipe-laden cars to the foot of penstock construction, where another of the Hetch Hetchy's inclines took over. The railroad's locomotive crane transferred pipe sections to a special standard-gauge flatcar which ran on a track alongside the penstock route. Operated by a 300 hp electric hoist, this tramway had a capacity of 25 tons. When a pipe section had taken its cable-car ride up Priest Hill, it was lifted from the car and delivered into place by a stiff-leg derrick, which rested on that portion of the penstock already secured. The derrick was moved along on skids also fastened to the pipes. A light trestle carried Moccasin Incline Tramway over the mainline railroad at Cavagnero Siding.

The City of San Francisco had acquired about 163 acres in the valley of Moccasin Creek, not far from Hughes Ranch and famous Hughes Station. The latter had been an important stagecoach junction in early days, where Big Oak Flat Road branched off, crossed a covered bridge and began the hated climb to the Priest Hotel and gold camps above. Now, in 1923-24, just across the way, a massive powerhouse building 225 feet long was to be erected. Designed by architect H. A. Minton in old Spanish Mission style with tile roof and wide arcades at the sides, the building would be so constructed that the south end could be extended for two additional generators when needed.

A "tent city" construction camp for 300 men was completed early in 1922 near the proposed powerhouse site, and preliminary work was begun: digging out the foundation area with steam shovel and 0-4-0 dinky locomotives, laying track, and receiving and storing materials. The Hetch Hetchy Railroad tracks were laid right into the powerhouse to facilitate the installation of eight big Pelton water wheels and four 20,000 kv-a

Everything in readiness, March 1923. The 225-foot-long, Mission-style powerhouse is complete; penstock pipes have been anchored to the hillside and firmly riveted together; the power transmission line has been extended 98½ miles to the edge of San Francisco Bay. Here workmen are stringing power cables on No. 2 tower.

generators. A 135-ton capacity permanent overhead electric crane was set up inside the building and used to lift machinery off the railroad cars and drop it in place. Owing to the almost ideal topography of the site, it was possible to install all transformers, high-voltage switches and busses just outside and in back of the powerhouse. The installation was completed under the direction of Andy Johns and Orland and Andy Townsend. Moccasin was all set to go by June of 1925.

On Sunday, June 7, the first water from Hetch Hetchy flowed into the power tunnel at Priest, down the penstock and through the 11-inch nozzles at the powerhouse. Various tests and drying out of windings were conducted for the next few days until everything was ready for operation at full voltage. At 6:58 on the morning of June 30 a powerhouse man was making one final test and, according to the City's report, opened a main valve without opening the by-pass, thus setting up a pressure surge in No. 3 penstock. This was communicated to No. 4, bursting a section of the latter at a defective weld about 1000 feet above the powerhouse. A section of pipe in No. 1 let go likewise. Timekeeper Jim Neel looked up beyond the railroad and saw a wall of water start down the mountainside. Tents and wooden houses to the left of the penstock were wiped out entirely; autos were jammed against the spur-line railroad trestle; trucks were shoved out of the way. Frank Cummings had to climb a tree to avoid being washed down the canyon. It took nearly half an hour to close a temporary manual shutoff up above, but the damage was all done in less than five minutes, leaving the powerhouse basement filled 15 feet deep with mud and boulders!

Nearly six weeks were needed to clean up and dry everything out; so the powerhouse didn't start producing revenue until mid-August. The water used to develop this power income was allowed to flow out of Moccasin Powerhouse at elevation 935, down Moccasin Creek and into the Tuolumne at Stevens Bar. There was as yet no aqueduct line to carry drinking water westward to the City. But within just a few months work was started across

One of the burst penstock pipes after the disastrous failure of June 30, 1925. Homes were wiped out; mud and rocks filled the powerhouse basement far below, and a portion of Hetch Hetchy Railroad was left hanging in mid-air. *(—Jim and Marie Neel)*

Officials in charge of Hetch Hetchy electrical installations. Left to right, brothers Andy and Orland Townsend and Andy Johns, at the Moccasin switchyard, 1925. *(—Frank Cummings)*

A by-pass conduit was necessary to carry possibly contaminated waters of Moccasin Creek beneath the reservoir below the powerhouse. This conduit is shown here during construction in 1929. Beyond is a stockpile of gravel crushed from Foothill Tunnel granite, to be used later in tunnel lining. Standing by are two small gasoline locos, two-foot gauge. *(—City of S.F.)*

The completed reservoir at Moccasin is shown in this view looking downstream in 1936, after all the works had been completed. Powerhouse is to far right and Moccasin Dam in center distance. In foreground is Moccasin Diversion Dam, which was built to channel waters of Moccasin Creek into the conduit shown and carry them under and beyond the drinking water in the reservoir. *(–City of S.F.)*

the small valley on the Foothill Division Tunnel. This would carry water another 16 miles on its journey and deposit it in pipelines for the long trip across the San Joaquin Valley. When the Foothill Tunnel was completed, an 850-foot dam 55 feet high was put in below Moccasin Powerhouse to form an afterbay leading to this tunnel.

There was to be a problem at Moccasin caused by water flowing down Moccasin Creek and mixing with water from Hetch Hetchy. This couldn't be tolerated during heavy winter runoffs, as the local water was subject to contamination from cattle ranges and operating gold mines upstream. For this reason, a few years later a sort of water by-pass tunnel nearly 3000 feet long was laid underneath Moccasin Reservoir and a smaller diversion dam built upstream. Thus the dirty water would pass under and not mix with the pure Hetchy stuff. The two dams here, by the way, were sixth and seventh in the original construction phase of the project.

While the powerhouse was under construction, Hetch Hetchy forces built six cottages nearby for operators and other employees. A garage and a schoolhouse were constructed, also a large clubhouse to provide quarters for single men. The clubhouse had a big central living room on the ground floor which served as a sort of community center. As the years went by, Moccasin was destined to become almost a town in itself, with scores of dwellings, offices and shops – headquarters for construction and maintenance of the entire Hetch Hetchy system. The superintendent at Moccasin now acts as unofficial "mayor" of

Electrical generating machinery was installed inside Moccasin Powerhouse in late 1924 and the first part of 1925. Hetch Hetchy Railroad tracks entered the south end of the building in order that the heavy armatures, shafts and other pieces could be brought right under the permanent overhead crane. The man to left has been identified as Frank Reese, while "Sealskin Pete" Peterson stands on stepladder. Below is one of the shafts supplied by Pelton Water Wheel Co.

The completed plant at Moccasin is shown in this 1959 view from the powerline tower, looking eastward toward Hetch Hetchy. To left of the powerhouse is the Club House, originally constructed as a boardinghouse for single men. Railroad tracks have been removed and replaced with shops and parking spaces to the right. Big Oak Flat Road climbs the mountain from upper left; it is now Highway 120, an all-year route to Yosemite. *(–S.F. P.U.C.)*

this pleasant, tree-shaded settlement about twenty miles south of Sonora.

Electricity generated at Moccasin had to be carried to San Francisco – at least, that was the original plan as agreed in the Raker Act. A transmission line was started westward in 1923, a double line across the hills and valleys almost exactly on the aqueduct right of way. Early in 1925 this line was ready: 98½ miles to the edge of San Francisco Bay at Newark, two circuits carried on a string of 506 steel towers. When the system was activated in August, everyone was delighted. Designed for a continuous load of 52,000 kw., it was able to carry 60,000 thanks to conservative planning and lower line loss than anticipated.

Attacks had been made on accounting methods used by the City during construction of the Hetch Hetchy Project. In the *Journal of Electricity and Western Industry* for June 1, 1922, writer Paul Eliel did a hatchet job in the best traditions of those who constantly aim to distort facts about public ownership of utilities. He was answered in the November 15 issue of the same journal by a lengthy letter that refuted all charges and allegations he had made and ended by stating that the accounting methods used on the project "are conspicuous examples of accounting methods that might profitably be studied by privately owned utilities." This almost-unique letter was signed by William Dolge, who was at the time Secretary of the State Board of Accountancy and President of the California Society of Certified Public Accountants.

However, now that Hetch Hetchy started bringing in some money in 1925, such attacks almost completely disappeared, leaving the grumbling and sniping to self-seeking politicians. There would still be underhanded attacks, eagerly reported in the press, before water actually flowed into San Francisco in 1934. But with the production of power the first tangible results were in hand, perhaps a little late but exactly as planned, and the taxpayers were able to see and understand that their Department of Engineering was doing a really good job for them. There was also a more subtle consideration which would be of increasing importance in California: power generated by fuels implies a drain on natural resources and resultant pollution, but hydroelectric energy is the essence of conservation.

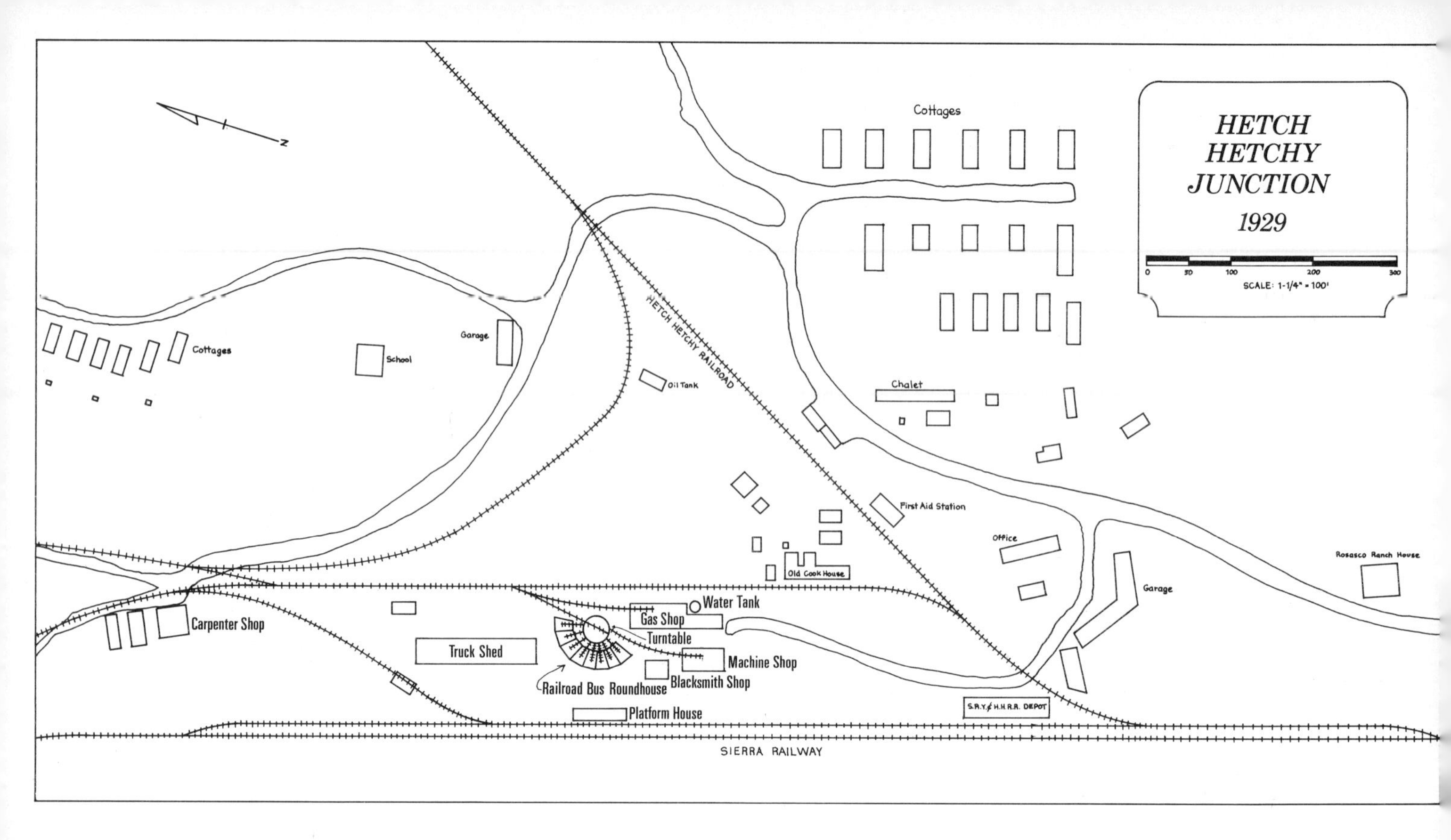

In 1925 headquarters for the construction work was moved from Groveland to Hetch Hetchy Junction, as the aqueduct line was stretched westward. The junction was in a narrow canyon, the only flat portion being that at the bottom of the map above. The upper legs of the railroad wye were both on steep grades. Foothill Tunnel line passed just to the left of the map. The bond below is part of the ten-million-dollar 1925 issue to finance construction of Foothill Tunnel and other works.

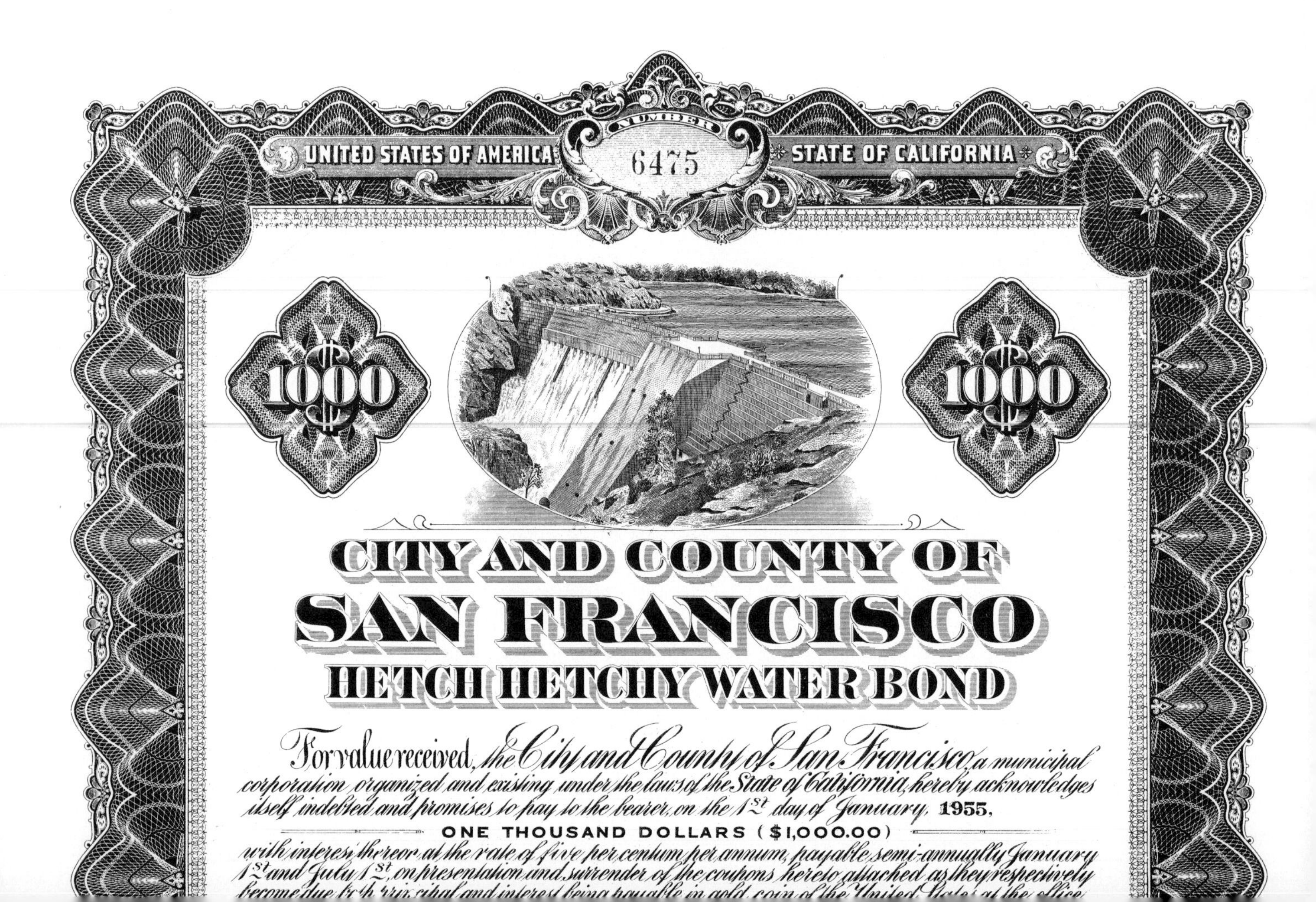
UNITED STATES OF AMERICA
NUMBER
6475
STATE OF CALIFORNIA
$1000
$1000
CITY AND COUNTY OF
SAN FRANCISCO
HETCH HETCHY WATER BOND
For value received, the City and County of San Francisco, a municipal corporation organized and existing under the laws of the State of California, hereby acknowledges itself indebted and promises to pay to the bearer, on the 1st day of January, 1955,
ONE THOUSAND DOLLARS ($1,000.00)
with interest thereon at the rate of five per centum per annum, payable semi-annually January 1st and July 1st, on presentation and surrender of the coupons hereto attached as they respectively become due, both principal and interest being payable in gold coin of the United States of the ...

18

WESTWARD BY TUNNEL AND PIPELINE

William "Red" Wanderer started with Hetch Hetchy on the Foothill Division and was later a resident engineer on San Joaquin Pipeline construction. Speaking of O'Shaughnessy, Red said, "The Chief was a grand old man . . ." who would walk each morning through the engineering department's drafting room at City Hall and would know what each man was working on. He'd go out and inspect every street job at least once a week and knew everyone on the jobs by name. Wanderer added that this never happened under the Chief's successors, which may account for a difference in results. And the Old Man was up in the mountains checking on his beloved Hetch Hetchy at least once every four or five weeks. Bill Helbush, retired senior engineer, said the Chief was contact man between the office and the field. He'd change clothes in the office closet down at City Hall and set out for the mountains with armloads of drawings, often stopping overnight at Byron for the hot springs treatment en route to Tuolumne County.

After the clubhouse was built at Moccasin, the Chief usually spent the night there while on his rounds. Slim Jameson remembered that, although prunes weren't normally served in the dining room, when O'Shaughnessy was there he would personally dish out exactly seven prunes for each man at breakfast. His heart was in the job to the point that the men's health and welfare concerned him as well. And he had the heart to fight for Hetch Hetchy too, as he had to in 1924 when Moccasin Powerhouse was almost ready and work could soon begin on Foothill Division.

A $10,000,000 Hetch Hetchy bond issue was on the San Francisco ballot for October 7, 1924; this money was needed to construct a tunnel westward from Moccasin and start preliminary work in the Coast Range near Livermore. Approval of these bonds was recommended by everyone: work would stop without them and crews would be dispersed to the four winds. O'Shaughnessy was all over the City, speaking to clubs and church groups in the evening, to City employees during lunch hours, to businessmen and civic leaders. Another chief engineer at age 60 would have said "the hell with it" and just waited for the vote. But in the 1920s this man was the highest paid city engineer in the United States, and he deserved it. He went around and explained the plans, the progress to date, what needed to be done and why more money was needed. He told the people, "Give me the ten million now and I'll be back in a couple of years for twenty-three million more to finish the job." The bonds passed by an overwhelming vote and the work went on.

From Moccasin Reservoir, Hetch Hetchy Aqueduct continues westward in an entirely enclosed system for the next 116 miles to Crystal Springs Reservoir on the City's doorstep. The first section is a 16-mile tunnel through the foothills on the east side of the San Joaquin Valley to the beginning of the great valley pipeline at Oakdale Portal, south of Knight's Ferry. To start Foothill Division work, the first step was to close construction headquarters at Groveland and move shops, equipment, offices and dwellings to a narrow, sun-baked canyon at Hetch Hetchy Junction, the starting point of the railroad. This was an ideal location for supply and communications and it would be the site of one of the shafts for tunnel construction.

Work on Foothill Tunnel, running 16 miles westward from Moccasin, started in 1926. The upper view, taken near Moccasin Portal, shows the crushing plant which converted tunnel rock into gravel and sand for later use in concreting parts of the bore. Four miles west of Moccasin was Brown Adit, from which the tunnel work went on in both directions. To the right is a view of the shop area at tunnel portal, and below some of the camp bunkhouses are shown. *(—Bottom, City of S.F.)*

Shown at top is the portal at Brown Adit, fresh-air pipes entering the tunnels from blower house to right. The two-foot-gauge railroad tracks, used to carry muck from the digging faces, connected with a novel passenger and freight carrier named Brown Adit Tramway. It was over 1½ miles long, running out to the river canyon at Red Mountain Bar in order to eliminate use of a terrible truck road from Moccasin. The locomotive is a six-ton gasoline job supplied by Plymouth Mfg. Co. Passenger to left is Construction Engineer Lou McAtee; the others have been tentatively identified as George Laveroni and Alvin Gray. *(—Upper, City of S. F.)*

Tunnel face in Moccasin heading in February 1927. Men standing on the platform are drilling the upper set of a round of blasting holes while mucking goes on below.*(—City of S.F.)*

At each tunnel camp two men were kept constantly at work sharpening the drills for digging into tunnel faces. Various lengths were used in order to prepare holes for blasting, and in the hard granite frequent sharpening was necessary. The helper to left is heating the steel while the other man does actual sharpening in the Ingersoll-Rand machine to right. Brown Adit Camp, January 1928. *(–City of S.F.)*

Practically everything was moved out of Groveland in 1925. The water supply system was shut down when townspeople couldn't get together on taking over operation from the City. Only the hospital was left, but with no trains running past the front door all was silent at last. And old Groveland was quiet again, its boisterous decade at an end.

The railroad that took up its new headquarters at the junction was considerably changed; in fact, it listed only 13 permanent employees at the time. Shay No. 6 and both Mikados were put up for sale, only engine No. 5 and a few flatcars being retained for possible heavy loads. Passenger track buses Nos. 19, 20, 23 and 24 were kept, as were two track trucks, one being former bus No. 21. A small roundhouse and light turntable were installed for the gas equipment, Dud Snider being in charge. Five passenger coaches became superfluous, as did about 35 freight and work cars and operating equipment: boxcars, flats, ballast and gondola cars, the snowplow, the steamshovel and two cabooses. The engines were soon sold and a few of the well-used cars, but many of the latter were still rotting away at Moccasin when the final end came in 1950.

There was no town at Hetch Hetchy Junction as there had been at Groveland – only Rosasco's ranch house down the canyon and Sierra Railway trains passing by from time to time. But with the moved structures and new ones, headquarters was rapidly re-established here under Lloyd McAfee. They were ready for business on November 27, 1925. Shops, warehouses and offices were down near the usually dry creekbed, with dwellings rising up the hillside among oaks and rattlesnakes. A schoolhouse was part of the camp and a good number of children attended classes conducted by Margaret Golub Snider.

The Foothill aqueduct was surveyed to pass almost directly under this settlement. Construction camps were located at Moccasin Portal, Brown Adit (4 miles to the west), Red Mountain Bar (at the Tuolumne River Canyon, about 1½ miles), Pedro Adit (¼ mile west of the canyon), Hetch Hetchy Junction Shaft (4½ miles), Rock River Shaft (3 miles) and Oakdale Portal (3 miles). To get these camps established and full work under way, new roads had to be built, extending from the junction to Rock River, and from there on to Oakdale Portal. A steep and narrow wagon road from Moccasin to Brown by way of Marsh's Flat was improved temporarily while means were sought to by-pass it. At the canyon, the railroad serviced Pedro Adit with a spur and Red Mountain Bar with a cable incline. Finally, power and phones were brought in and a pipeline installed to carry a pure water supply from Priest Reservoir to each of the dry Foothill camps. This pipe was laid along the rail line as far as the junction, thence on the surface above the aqueduct right of way. A branch was carried to Brown Adit, using cable suspension over the deep river canyon.

Early in 1926 tunnel work began at Pedro Adit. Soon miners were drilling and blasting at ten faces: two from the portals, 16 miles apart; four from the two shafts, at the junction and Rock

At Red Mountain Bar, about 5½ miles west of Moccasin, Hetch Hetchy Aqueduct had to cross Tuolumne River Canyon by means of a steel-pipe siphon. The pipe itself was cemented into the riverbed ahead of other Foothill Division work, since it was in an area to be covered with water upon completion of the first Don Pedro Dam in 1923. The view below shows incline railway dropping down from Hetch Hetchy Railroad to the river in 1920. Seven years later, when work was in progress on the tunnel here, the reservoir water covered the siphon, as shown to right. A half-mile-long Lidgerwood cableway spans the canyon, carrying men and materials between the railroad and Brown Adit Tramway. Near bottom of picture can be seen the platform en route with two passengers. Both views look westward. *(—Both, City of S.F.)*

Upon completion in 1923 the inverted siphon, a 9½-foot steel pipe, was embedded in concrete under the Tuolumne River (later Don Pedro Reservoir). View looking eastward toward Moccasin. *(—City of S.F.)*

River; and two each from the adits at Brown and Pedro. Five miles west of Moccasin the tunnel was broken by the river canyon at Red Mountain Bar. Here the half-mile crossing had been conquered in 1923 by the inverted siphon, a 9½-foot steel pipe lined with and embedded in concrete. This work had to be done so much earlier because in that year the original Don Pedro Dam had been dedicated and began backing up water over the site. For the same reason, the HHRR wooden trestle at Six Bit Gulch, a mile or so upstream, was replaced in 1923 with a higher steel bridge, 585 feet long, on tall concrete piers.

Most of the tunnel construction equipment used at Foothill had been salvaged from Mountain Division work up above: compressors, pipes, blowers, Myers-Whalley and Conway electric mucking machines, Webb and Hackley dump cars, and four-, five- and six-ton storage-battery locomotives.

The latter had proven themselves over the years since 1919, when there had been some doubt as to their effectiveness in contrast to trolley operation. It was found that power consumption compared favorably with that of overhead trolley equipment, and batteries provided an additional safety factor. Efficient and relatively fast running was assured by equipping all cars with roller bearings and installing 30-pound rail on wooden ties with rock ballast throughout. Fifty-ton trains made only 3.9 mph against the tunnel grade — but upgrade was only half of the running. Battery recharging time averaged about twice running time, or 16 hours out of 24, but this worked out very well with the number of loads to be removed.

Steel pipe in 24-foot sections, each weighing 12 to 16 tons, was transferred from Hetch Hetchy Railroad at Red Mountain Bar, onto the standard-gauge tramway which dropped it down into the Tuolumne River gorge to form the siphon shown on opposite page. *(—City of S.F.)*

"Motor" operators were paid $5.50 a day in the Foothill tunnel; mucking machine men got $6. Outside haulage equipment was employed at three of the camps to distribute spoil to the dumps. Each had one or two gasoline locomotives and western dump cars. An eight-ton Plymouth performed these duties at Hetch Hetchy Junction Shaft, mainly disposing of rock that had been deposited in bins at the surface. Purchased in 1928, this was the first of the three Plymouth locomotives that would stay with HHRR until the end.

San Francisco's day-labor employees, under direct charge of the City Engineer, installed all necessary appurtenances at the camps and then drove 1000 feet of tunnel in each heading so that prospective contractors would have an idea of the conditions. About half of the tunneling was turned over to the two successful bidders, A. Guthrie & Co. and Connolly & De Luca, and the unusual decision was made to continue with City forces on the other half.

The road to Brown was a terror, so during the summer of 1926 crews built a 24-inch-gauge tramway over a mile and a half long from the adit out westward to the canyon. It connected directly with the track into Brown Adit, though motive power on the new line would be a homemade-looking six-ton Plymouth gas locomotive. When the tramway tracks reached Red Mountain Bar on August 11, workmen could see lots of activity across the reservoir, around Pedro Adit and the railroad siding, but there was no way to get across until a 2300-foot Lidgerwood cableway began operating on September 13. This overhead line carried peo-

For a short distance immediately west of Red Mountain Bar siphon, the aqueduct tunnel was in circular form, as far as Pedro Adit. The rail-mounted concrete mixer and placing gun are shown being set up for tunnel lining work in 1929.

Westward from Pedro Adit the tunnel resumed the horseshoe shape. Here at Pedro are shown collapsible metal tunnel lining forms, ready to be transported inside by special carrier cars on the narrow-gauge tracks. *(–City of S.F.)*

The completed short section of circular tunnel at Red Mountain Bar west portal as the forms were being removed.

Tuolumne County provided a school and teacher for the camp at Hetch Hetchy Junction. All pupils, of course, were children of the City's employees, and the teacher (center, back row) was Mrs. Margaret Snider. The year was 1927. *(–Irene Magee)*

ple, supplies, machinery and parts on a two-minute ride from the mainline rail siding to a wooden platform alongside the tramway. The cableway's capacity was six tons, and this was fortunate because the first thing across was the six-ton locomotive. Under this load, it is reported, the cable sagged 149 feet, but the two wooden towers, 61 and 85 feet tall, that had been designed by Owen Ellis showed no sign of strain.

As the tunnel faces pressed ahead, a mighty rivalry sprang up between City employees and the contractors' men, which resulted in exceptional progress. Three times the Chief's forces broke existing records for speed in tunnel driving, establishing a final record of 803 feet of advance in the month of September 1927. An interesting contrast: costs were $35.53 per foot by City crews, as against $40.49 by the contractors. There was an even greater difference later in placing the concrete lining. Fine men were employed at every level on the Hetchy job and the spirit generated was something rarely encountered today; it comes to light, however, whenever a Hetch Hetchy oldtimer starts talking about the job and the people who did it.

The last section of the Foothill Tunnel was completed December 6, 1928, when Moccasin to Brown was holed through, with the center line out just half a foot! Then concrete lining began on 45% of the distance, the balance being solid granite. Three cement plants were required for this work, the first at Moccasin Portal, where a crusher had been working since September 1927 on suitable rock as it was brought out of the tunnel. From this plant tunnel lining proceeded all the way to Red Mountain Bar. A second concrete plant was set up above the shaft at Hetch Hetchy Junction, to handle all lining from Red Mountain Bar to a point halfway to Rock River. Here all aggregate had to be purchased, coming in over Sierra Railway spur tracks leading directly to the shaft. At Rock River the contractor operated a third concrete plant, using tunnel spoil for aggregates.

Methods and equipment for the lining job were the same as had been used on the Mountain Division Tunnel. A "jumbo" on wheels contained all equipment necessary for handling concrete from delivery of dry batches to placing inside the forms: an incline, a conveyor, a concrete mixer and a pneumatic gun – all operated by compressed air. After the tunnel was lined, the siphon at Red Mountain Bar had to be extended up to the portals on both sides of the canyon, since only the underwater portion was completed in 1923. Missing sections were added in 1932 as time for finishing of the entire aqueduct drew near.

Cleaning up the details, after completion of Foothill Division, went on for several years, especially as concerned claims of various farmers and

Foothill Tunnel passed underneath Hetch Hetchy Junction. Shown at left are the shops and shaft headworks, with dump of tunnel rock in foreground. This 1927 view looks eastward from the railroad yard area at the junction.

Looking westward at Hetch Hetchy Junction from the tunnel shaft. Material yard in foreground; Sierra Railway at far right; line of seldom-used HHRR flats and boxcars left of center.

Three miles west of the junction was another tunnel entrance, Rock River Shaft. Here is the camp layout in February 1927. *(–Three photos, City of S.F.)*

By the end of 1931 work was going on all across San Joaquin Valley as City forces prepared to lay a 47-mile pipeline. The machine shown above is a Buckeye Trencher, largest in the West at that time, digging out a trench between Riverbank and the San Joaquin River. Right, the river crossing is prepared where the pipe would be encased in concrete below the navigable streambed at an elevation 1½ feet below sea level. *(—Upper, City of S.F.)*

The San Joaquin Pipeline route took Hetch Hetchy Aqueduct underneath several railroad lines. Here work goes on below the Santa Fe main line near Riverbank in 1931.

cattle raisers for losses of water, allegedly because the aqueduct passed under their lands. This time, however, surveys had been made for years before the work started so that many outlandish claims could be contested and thrown out. The county roads and state highways used for the City's hauling were maintained and repaired so that they were left "at least as good as, and generally better" than before.

The Hetch Hetchy Railroad continued to operate out of junction headquarters, but on the very restricted basis that followed removal from Groveland. Steam engine No. 5 had been in service during May and June 1927, hauling materials for the Moccasin diversion tunnel, new cottages near the powerhouse and a sewer and water supply system for the Mather tourist camp. Afterwards, it was determined that the small Plymouth locomotive working at Hetch Hetchy Junction was able to haul one carload of sand or cement as far as Moccasin, so they put it to work on the main line and locked the 5-spot away for good.

Headquarters at the junction was closed on March 19, 1930, and most of the staff was transferred to the Coast Range Division, already under way. Leased property was thoroughly cleared off, several cottages and the truck garage being moved to Moccasin and set up there for use by Power Division employees. When a small advance party of railfans ventured down the rocky road to Hetch Hetchy Junction in 1937, all they saw left was derelict track bus No. 20, sitting in the flimsy open "roundhouse" with side curtains flapping in the breeze. Not far away, of course, the main line was busy with Sierra Railway equipment carrying materials to Mather for a dam-raising project, about which more later. From 1930 on Moccasin was headquarters for the Hetch Hetchy Railroad. Dispatching was handled by the powerhouse operator who happened to be on duty, as the railroad was maintained to serve the Power Division, providing the only transportation for maintenance of way and telephone lines during winter months when highways and roads were impassable.

So construction work on the aqueduct moved westward again and Lou McAtee was put in charge of building the San Joaquin Valley Pipeline, operating from an office at Tracy. This great pipe was 47½ miles long, from Oakdale Portal at elevation 747, to Tesla Portal, elevation 399, about seven miles south of Tracy. At the lowest point, passing beneath the San Joaquin River, the line was 1.4 feet below sea level. In initial construction only a single pipe was put in place, with a capacity of 60 m.g.d. It was imperative to get this much water going by mid-1932, based on pro-

In October 1931 Youdall Construction Company's giant trenching machine reached the San Joaquin River, which was crossed on temporary trestlework, and then continued working its way westward toward the Coast Range.

City engineers and contractor officials inspected the concrete pipe-wrapping plant at Modesto, October 31, 1931. Second from left, Jim Foss. The fourth man, of course, is Chief O'Shaughnessy with dark suit and carrying a cane (he was 67 at the time). The men to right are Meadowbrook and Jim Turner, engineers from the city office in San Francisco. *(—City of S.F.)*

Two additional Hetch Hetchy pipelines have been laid across the San Joaquin Valley since that first one of 1931-32. This view shows the third, a 78-inch line, being placed in 1966. The City's power transmission line follows the same right of way. *(—S.F. P.U.C.)*

The riveted Hetch Hetchy pipeline No. 1 proceeds across the San Joaquin Valley in October 1931. The upper view shows field joints being riveted, while below left sections of pipe are lowered into the trench. *(—Both, City of S.F.)*

The pipeline was completed and all openings west of Moccasin were sealed during 1932. Below is the steel inspection door at Pedro Adit in 1964. All that remains of the tunnel camp are a few foundations.

The third pipeline, placed in 1966, used bigger pipe, different equipment and modern methods. But all three lines lie side by side in the 100-foot right of way across the valley, and there's room for a fourth, which will bring the Hetch Hetchy Aqueduct up to designed capacity. *(–S.F. P.U.C.)*

jections following a succession of dry seasons. A contract was awarded in May 1931 to the Youdall Construction Co. to build the line of welded steel pipe, laid eight to nine feet deep.

The ditch from Riverbank west was dug with a monster trenching machine, said to be the largest ever operated to that time in the West. East of Riverbank toward Oakdale Portal the ground was cemented gravel and had to be "shot" every twelve feet. Engineer Red Wanderer said the top nine miles of this line was the first really big welded job anywhere and was completely successful, passing all tests. Varying in diameter from 57 to 72 inches, sections of the pipeline were coated with hot asphalt and wrapped in asphaltic felt. At crossings of the San Joaquin River and Elliott Cut, both classed as navigable streams, pipe was depressed below the streambeds and encased in reinforced concrete. The line was completed in a year, joining the mammoth Coast Range Tunnel at Tesla Portal.

To celebrate completion of that first pipeline back in 1932, there was a dinner in Oakdale, with guests from Modesto, Tracy and San Francisco. As Chief O'Shaughnessy noted in a speech at the time, "It is significant of the change of feeling of the people of San Joaquin Valley in the past 15 years from one of bitter antagonism to present friendly co-operation."

Hetch Hetchy has a 100-foot right of way across the great valley, with power transmission line above and room for four pipelines underground, so designed that additional pipes could be installed as more water was required. No change would be needed in tunnels at either end: they were built with sufficient capacity for all four pipes. Thus, San Joaquin Pipeline No. 2, with an inside diameter of 61 inches, was completed in 1953; the third, 78 inches, went into service at the end of the 1960s. Capacity of the three is 295 million gallons per day, so a fourth can be added when needed.

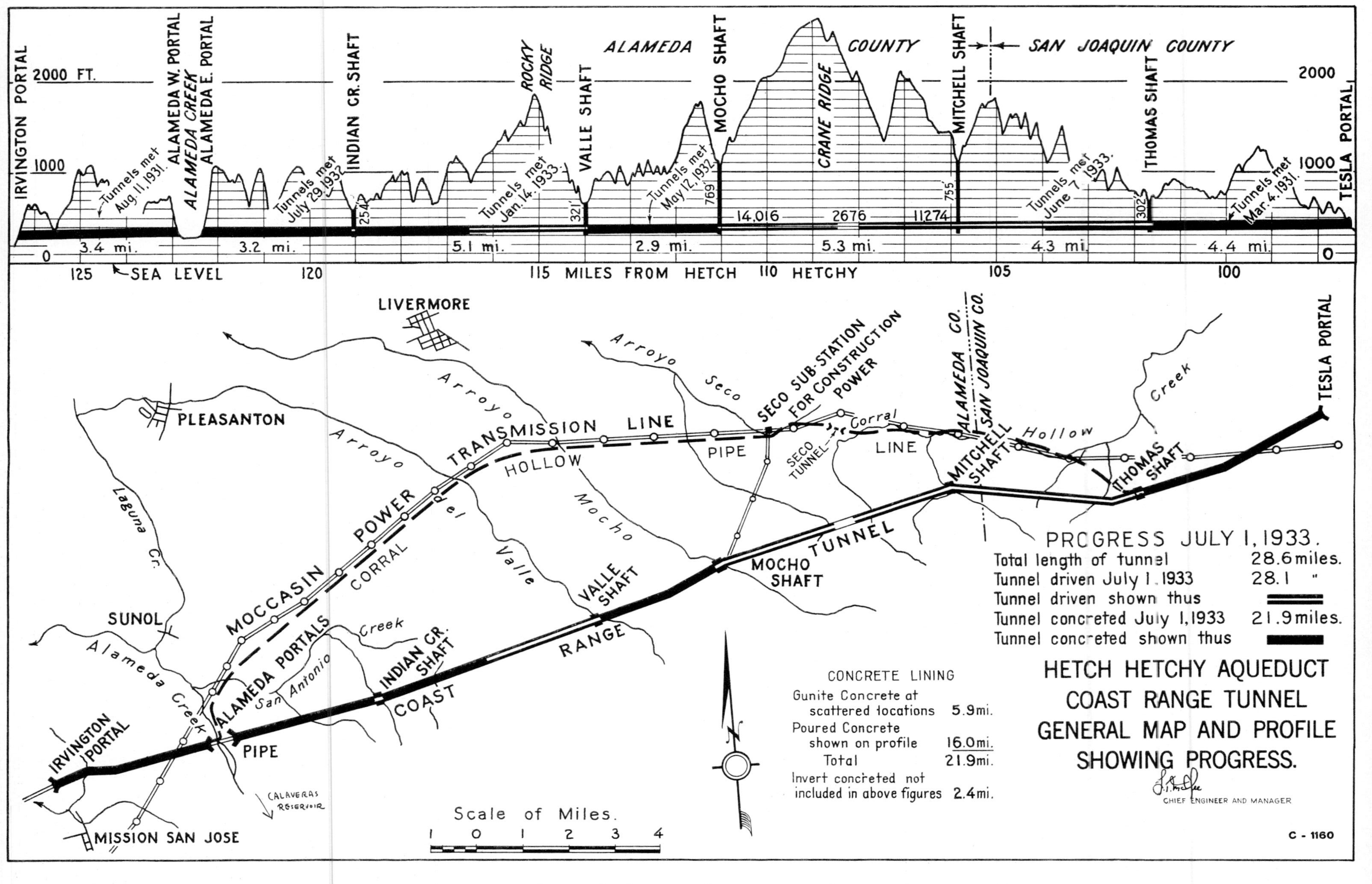

The Coast Range Tunnel, stretching nearly 29 miles from Tesla Portal, near Tracy, to Irvington Portal, near San Francisco Bay, was completed about a year after this 1933 drawing was made. Note that the Corral Hollow (Emergency) Pipe Line follows fairly closely the route of the City's power transmission line.

19

LONGEST TUNNEL IN THE WORLD

Hetch Hetchy Aqueduct's Coast Range Division extends from Tesla Portal, the end of San Joaquin Pipeline, westerly to Irvington Portal near Mission San Jose in Alameda County. Starting at elevation 399, the tunnel drops 83 feet in its 29-mile length, finishing at 316. There are actually three parts: a 25-mile tunnel, a 3½-mile tunnel and a half-mile pipeline joining the two tunnels across a valley at Alameda Creek. Survey and location work was done by engineer Max Bartell, who had made a geological study of the entire area with particular emphasis on earthquake faults. Bart usually worked out of the City Engineer's office, a sort of back-room planner, although Jack Best, one of the most admired engineers on the job, stated in a 1964 interview that Max Bartell was really "Mister Hetch Hetchy."

This new tunnel was to be considerably flatter than the mountain tunnels already completed; it dropped only three feet per mile. It would need to be considerably larger to carry that 400 m.g.d. coming from Hetch Hetchy. It was decided that, rather than construct one 13-foot tunnel, it would be more economical to build one to 10′6″ at that time to carry about half the flow, and another in 25 years or so when more water would be needed. The savings in cost, with future interest accumulations, would be more than sufficient to build a second, parallel tunnel at a future date. Many, many hours of consultation and discussion in Mike O'Shaughnessy's office were behind that decision. Undoubtedly, as much time had to be given to political implications and irresponsible newspaper reporting as to engineering and financial planning.

Another very important decision, which had to be made before the tunnel got under way in 1925, was whether to bother scooping out a record-size tunnel at all, or to simply pump water over the Coast Range hills in a pipe. That's what Oakland and her neighboring cities in the East Bay Municipal Utility District had elected to do, after finally pulling out of Hetch Hetchy. And, all else being equal, the pipe would save at least a year in construction time. However, at this point the Hetch Hetchy Project was in no rush; under the Raker Act it was not supposed to deliver water to San Francisco before 1932, at which time the resources of the Spring Valley Water Company would presumably be exhausted. They had plenty of time to dig the tunnel and in the long run it would be a lot cheaper. A pipe for only 60 m.g.d. would cost $8,000,000, compared to $16,000,000 for the gravity tunnel. But, per million gallons' capacity, it would be $266,500 for the pipe and only $65,180 for a tunnel. The small-capacity pipe, plus costs of pumping, would tie up nearly as much capital as a 250 m.g.d. tunnel, besides which, the pipeline would have to be added to repeatedly. O'Shaughnessy could easily have swung the decision to a pipeline and made himself look good, leaving future added costs for someone else to worry about. But his integrity wouldn't allow this, and his engineering staff was behind him unanimously. They held out for a quality job, despite hell and high water, newspapers and politicians. And the City got quality.

Politicians! San Francisco's Supervisors were worst. See how they started off: plans of the Department of Engineering called for starting Coast Range Tunnel work in the winter of 1925, but the budget was not approved by the Board of Supervisors until March 5, 1927. There was a

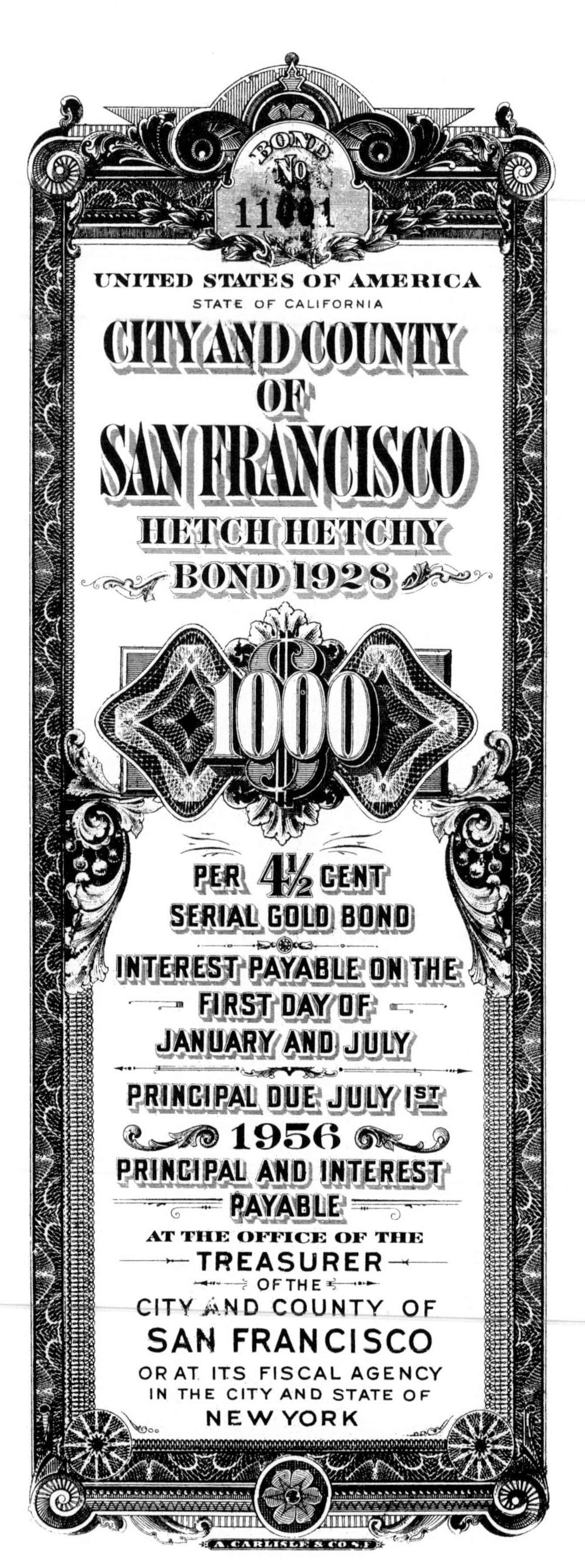

delay of 15 months, and later a good part of a year was lost as the result of an unfortunate accident. Then, when the job wasn't finished in 1932 as projected, O'Shaughnessy was ridiculed, derided, forced to submit endless answers to insulting questions and generally made the scapegoat. Although this man had been cheated twice in earlier jobs for the City and County of San Francisco, Rolph had begged him to return and do this last great job. In the end his hands were tied by self-seeking politicians and he was made to explain publicly why *he* hadn't got the work done! But the Old Man also had a loyal following and many admirers; in 1930 the University of San Francisco awarded him an honorary Doctor of Science degree.

Immediately upon approval of the budget the greatest engineering job of the Hetch Hetchy Project was under way. The town of Livermore was selected for the division headquarters. Some of the office and shop people at Hetch Hetchy Junction were transferred to Livermore in 1928; other workers came west as their jobs ended in Foothill Division. The hospital moved last, a new one being opened in Livermore in August 1930. Hetchy people noticed that the atmosphere here wasn't as friendly as they had experienced at Groveland. Livermore residents appeared clannish, so there was little socializing. Then too, the big city was only forty or so miles away now; there was always plenty to do there.

Roads were built, camps constructed, power lines and water systems laid in. Charlie Baird came along to patrol and keep his watchful eyes on the camps as he'd done so well up in the mountains. In the 25-mile tunnel section, five shafts were to be sunk, varying in depth from 254 to 823 feet. These, with the four portals, provided 14 faces from which material could be excavated. At the bottom of each shaft a crosscut had to be driven at right angles to the aqueduct, 95 feet to the center line of the first tunnel to be dug and 80 feet to the parallel tunnel planned for the future. From the north end of each crosscut the aqueduct tunnel would be driven in both directions. Work was started with City crews, all power coming from the system's Moccasin Powerhouse, taken off the transmission line at Seco Substation near Livermore.

Smiling faces at City Hall on December 2, 1929! Mayor "Sunny Jim" Rolph holds a check for four million dollars, just exchanged for that value in Hetch Hetchy bonds by Amadeo P. Giannini, right, President of the Bank of Italy (later Bank of America). Now the Coast Range Tunnel work could go on.

Tesla Portal was first camp in the Coast Range Tunnel. It sat at the western edge of San Joaquin Valley, south of Tracy, where the long pipeline ended and the tunnel started westward. It is shown to right in 1930, while below is Thomas Shaft and camp, 4½ miles to the west. *(—Both, City of S.F.)*

Mitchell Shaft was the next construction site. About four miles west of Thomas, the shaft went down a total of 804 feet. The top view shows shaft headframe and shop; next is a view of the camp area – bunkhouses and cookhouse – located in a narrow canyon, very hot in summertime. The view at bottom shows some of the warning signs facing tunnel men as they prepared to descend into the shaft. The rule against smoking was particularly difficult to enforce, according to several of the men who worked on this tunnel. *(–All, City of S.F.)*

Coast Range Tunnel "mine rescue squad" at Mitchell Shaft, August 1931. Except for a disaster at this location in 1930, the Hetch Hetchy had an outstanding tunnel safety record. *(—City of S.F.)*

The first camp was at Tesla Portal, just south of Tracy; here the 47½-mile cross-valley pipeline would be joined to the tunnel in a couple of years. Going westward toward the City, with mileages from Tesla, there were the following construction sites: shafts at Thomas (4.46 miles), Mitchell (8.66), Arroyo del Mocho (13.88), Valle (16.78), and Indian Creek (21.87), then portals at East Alameda (25 miles), West Alameda (25.7) and Irvington (29.14 miles, the total). In May 1927 the first shaft was started at Mocho. During the next two years all five shafts were completed, 12 of the 14 tunnel faces opened and four miles of tunnel driven. The work slowed down in fiscal year 1929-30, the start of the Great Depression, yet the year ended well with almost 2000 men on the job, a mile of tunnel driven in June alone and things running very smoothly underground.

As in the Hetchy's mountain tunnels, all inside haulage was handled by storage-battery locomotives, five-ton for the main line and four-ton for gathering or switching at the faces. Cars were two-yard side-dump, four to six in a train. When spoil had been conveyed to the shafts, it was dumped into rock pockets dropping 55 feet below the tunnel floor. There it was loaded by gravity into two-yard skips, which could be operated by counterbalance in two sections of the three-compartment shafts. Up on the surface rock trains were hauled to the various dumps by Ford and Plymouth gasoline locomotives; this was also done at portals where a long haul was involved. For ventilation there were the usual blowers, pushing fresh air from outside through 24-inch pipe to about 50 feet from each working face. The blowers were reversed for five minutes before each round was shot and for fifteen minutes after.

No two construction campsites were alike on the Coast Range job. Tesla was at the edge of the great flat Central Valley, reached by a new four-mile road from Carbona. Four and a half miles west was Thomas Shaft, 540 feet deep and reached by a two-mile branch of the road up Corral Hollow. The first holing-through in the Coast Range was in the tunnel between Tesla and Thomas, on March 4, 1931, and placing of concrete started immediately so that this first section by itself would be ready for use in connection with an emergency pipeline proposal.

Next along the line was the camp at Mitchell Shaft, also a couple of miles off the Corral Hollow road. The latter was a truck road laid by the City on six miles of the old abandoned railroad grade going into the hills from existing Moy station on the Western Pacific. Mitchell Shaft went down 804 feet. Mocho came next, five miles west, with the deepest shaft, 823 feet. It was eight and one-half miles south of Livermore, just a mile off Mines Road. Valle Shaft was just three miles away, as location of these shafts was not determined by distance but rather by the locations of canyons over the tunnel line. Another road out of Livermore approached Valle, with an old dirt road reconstructed for the last six miles.

Indian Creek was the final shaft, five miles west of Valle and only 304 feet deep. It was five miles east of the road running from Sunol to Cala-

Lunchtime 804 feet underground. At the crosscut, bottom of Mitchell Shaft, the battery locomotive serves as dining table while the digging crews from both faces get together for a mid-shift hot meal. Below, the entire crew at Mocho Shaft, including tunnel stiffs, loco motormen, muckers, timekeepers and shopmen. Tunnel foreman Pete Peterson is at far right, front row. The tunnel men, of course, included crews working in both directions from the bottom of the shaft. *(—Both, City of S.F.)*

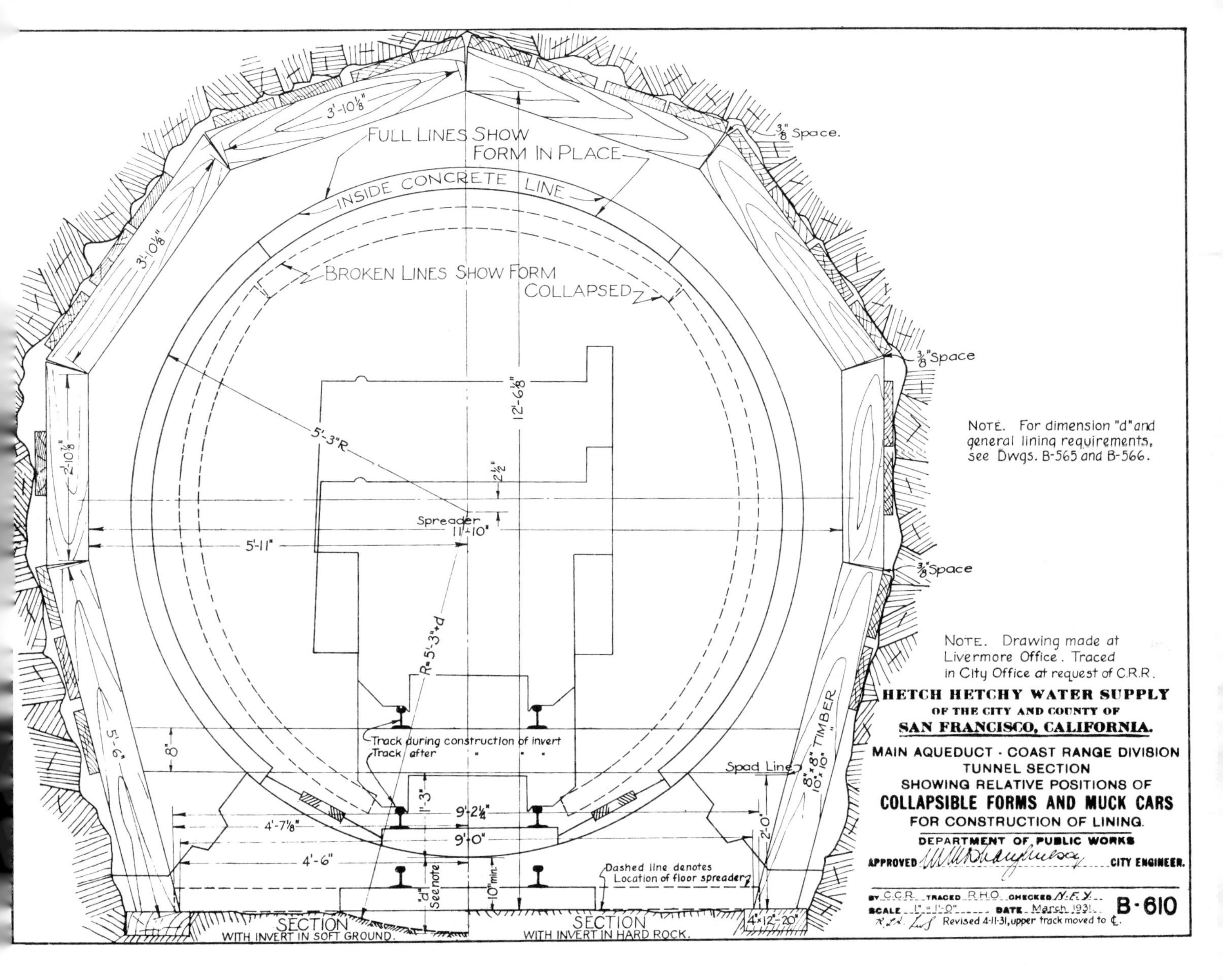

Coast Range Tunnel, being in softer and more treacherous ground than the mountain tunnels, had to be lined in circular form for the entire distance, as shown in drawing above. The photo below is a view of Mocho Shaft, placed in Arroyo del Mocho five miles west of Mitchell. This was the deepest shaft of all, taking 823 feet to get down to the bottom of the rock pocket under the tunnel. *(—City of S.F.)*

veras Dam. The tunnel in this area was in schist and gouge, which develop great pressure and require heavy and frequent timbering. The annual report of 1930 mentions repeated failures of timbers here, so bad that 1049 lineal feet of gunite concrete were placed during that year. There was soon more trouble, for on October 23, 1930, the west heading broke into a deposit of quicksand with heavy water flow. For a distance of 2500 feet back the descending tunnel was packed full of sand, which thereafter tapered off in depth to a few inches at the shaft crosscut. A second flow on November 18 carried sand into the east tunnel! To get back to the old face, a drift was first driven to drain the water, then widened to tunnel width, with gunite concrete placed to stabilize the roof. Eventually this area had to be completely gunited for a total length of 814 feet before work could proceed.

Three miles west of Indian Creek the aqueduct emerged at the valley of Alameda Creek, proceeded across this half-mile depression in a pipe, then entered a final tunnel section. On the west side of this valley another pipeline was met, this coming down from Calaveras Reservoir. Between the two portals on opposite sides of Alameda Creek a narrow-gauge industrial railroad was laid down, covering 4100 feet including 1100 on trestlework. A six-ton Plymouth gas locomotive was provided for use on the dumps, while the "main line" had a 16-ton Whitcomb diesel. The final construction camp for the tunnel was at Irvington Portal, where the aqueduct emerged from the hills and continued by a pipeline, which was already completed down to and across San Francisco Bay. Camp was laid out on the hillside near the settlement of Mission San Jose, with a lovely view of orchards and the distant Bay.

Project engineer Buddy Ryan got a phone call one night in July 1930 at his home in Livermore. It was from Carl Rankin, who headed the Coast Range Tunnel job, and he asked Ryan to hurry out to Mitchell and investigate a bad accident. At Mitchell everyone was standing around the top of the shaft; one hoist cable was missing. A workman had taken three boxes of powder and one of fuses into the bottom of the bucket and it had started slipping downward, unnoticed by the hoistman, who was distracted by the noise of another bucket coming up. The loose one fell 800 feet to the bottom, pulling four tons of cable down on top of it. Every bone in the worker's body was broken, but the powder didn't explode. Strangely enough, Ryan related, if it had gone off, the terrible accident of a day later wouldn't have happened.

During the night of July 17 there was an explosion of methane gas in the eastward heading at Mitchell, only 900 feet from the shaft. Twelve men — the entire crew — were killed, while 21 others working in the crosscut and westward made their way up the shaft to safety. Again Rankin phoned Ryan and told him to hurry over and investigate, but the man who went first into the smoking pit to see about rescue was electrician Jim Graham, member of the four-man rescue crew. As they started down the shaft, wearing Gibbs apparatus and face masks, Doc Degnan advised them, "Don't waste too much time; by now their brain cells are gone." And he added that if their nosepieces came off, they too were dead! All the men were frightened, so at the bottom Graham asked rescue captain Don MacDougall if he could go ahead while they waited, but to come and get him in a half-hour.

Five hundred feet along the tunnel he came to the first man, dead. Farther on he found the motor turned to one side with the motorman's body hanging out; beyond was what was left of the mucking machine operator, hanging from his seat on the other side. Graham had to crawl under one of the suspended bodies to get near the tunnel face. Right behind him came the rest of the crew and he breathed a prayer of thanks that they'd followed him. All of the working shift was dead. Graham noted that there were later rumors and tales that not all the crew had died in the blast, that ventilation pipes had collapsed and they had suffocated, and so on. This wasn't so, said Graham many years later; he said his report was the basis of the official report sent to state mine officials.

Ironically, there was a new air-vent line under construction and it would have been completed and operating four hours after this accident occurred, according to Jim Graham. It might have

The crosscut area at bottom of Mocho Shaft, looking north toward the actual tunnel. Dump cars in background are within the tunnel line. Pipes overhead form the ventilation system; their joints were made slightly flexible to avoid breakage in this earthquake territory. In the crosscut areas repair work was done, batteries charged on the locomotives and limited supplies maintained. *(—City of S.F.)*

Officials at top of Mocho, January 1934. Left to right: Dolph Wehner, safety engineer; Jackson Carle, City public relations man; Lloyd McAfee, Chief Engineer of Hetch Hetchy (Mr. O'Shaughnessy had been shoved aside by this time); Carl Rankin, project engineer on the Coast Range job; and tunnel foreman, Pete Peterson.

An early view (1928) of the crosscut at Indian Creek Shaft, looking north toward the tunnel line. The area is being prepared for the start of tunnel excavating in both directions from this point. *(—City of S.F.)*

In the valley of Alameda Creek, below Calaveras Reservoir, the Coast Range Tunnel was broken and the sections joined by an underground pipeline. This is the east portal and a portion of the ¾-mile tramway linking it with the west portal, February 1930. *(–City of S.F.)*

An emergency pipeline was rushed to completion between Newark and San Lorenzo in order to bring East Bay water to San Francisco during an unprecedented period of drought. Officials at the dedication on February 25, 1931, included Chief O'Shaughnessy (with watch chain). Supervisor J. Emmett Hayden is to his left. Third from right is George Pardee, head of East Bay M.U.D., then Mayor Rolph and Nels Eckart, head of San Francisco Water Dept.

Max Bartell, hydraulic engineer, was involved in Hetch Hetchy from the surveys of 1908 to his retirement in 1949. Bartell was called "Mister Hetch Hetchy" among the construction engineers on the project. It was his job to construct the Corral Hollow emergency pipeline (shown opposite). Later, he engineered tolerance, then understanding, and finally full co-operation between the irrigation districts and the City.

The Corral Hollow emergency pipeline was planned and built during the drought period in 1931-32 to bring Hetch Hetchy water to the City ahead of completion of Coast Range Tunnel. These photos were taken by engineer Max Bartell, who headed the project. Pipes were placed just below the surface so they could be removed and used elsewhere in later years.

Valle Shaft was located in Arroyo del Valle, south of Livermore and three miles west of Mocho. It was just over 300 feet down to the tunnel line at this point. *(–City of S.F.)*

The 24-mile Corral Hollow emergency pipeline was completed in just six months, ready to provide Hetch Hetchy water to San Francisco two years ahead of tunnel completion. View looking west in May 1932, with power transmission line to right. *(–City of S.F.)*

This White Motor Company ad of 1934 shows Hetch Hetchy Water Supply truck No. 115 at Livermore. It was a five-ton White, purchased by the City in 1914 and used continuously on all divisions of the project. Mechanics remembered that these trucks had wonderfully durable engines.

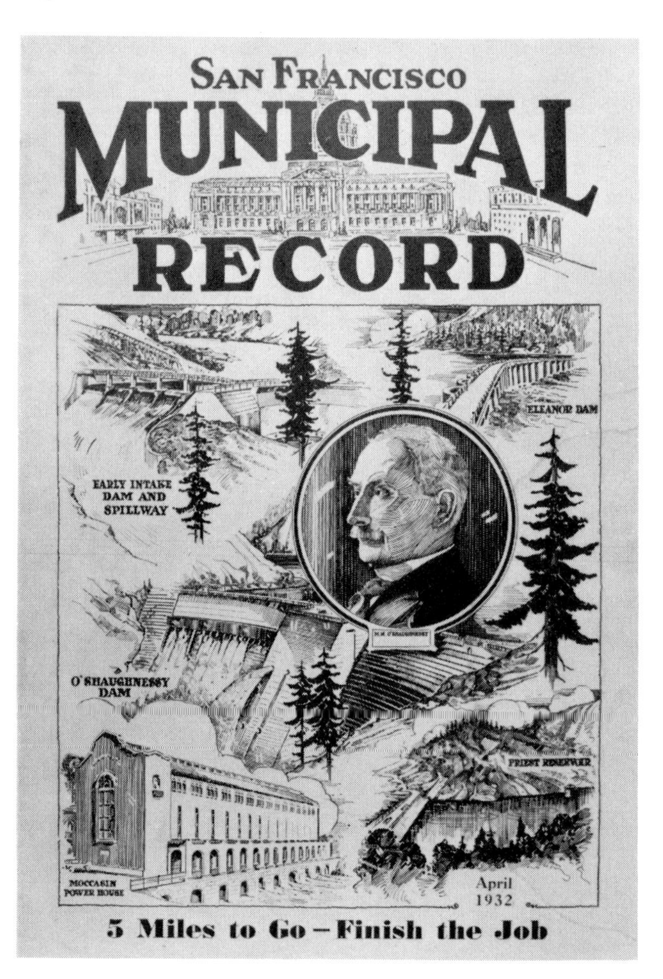

April 1932 issue of the *San Francisco Municipal Record* featured an appeal to voters for approval of the final bond issue, of $6,500,000, to complete the aqueduct and bring water to the City. The Chief Engineer, Mike O'Shaughnessy, is shown surrounded by four of the dams completed and Moccasin Powerhouse.

prevented the explosion. But what was the cause? Graham, who many years later retired as Chief Electrical Inspector for San Francisco, thought it happened when one of the miners "flipped" his light in the hole, instead of going to a fresh-air station. Several others were convinced that it was simply a case of a man sneaking a smoke, even though matches and all smoking materials were banned and there were frequent surprise inspections. The presence of dreaded methane gas did not seem to deter the smuggling of smokes – up to the time of this accident. *Engineering News-Record*, the authoritative weekly, listed three possible causes in its next issue: a broken light globe (though they had double guards), sparks from a motor or an infraction of the no-smoking rule.

Safety measures in common use in similar mining work had been followed, and yet the explosion occurred. Many investigations were made and, while the City was exonerated from blame at a coroner's inquest, additional safety measures were ordered by State of California industrial safety engineers. "Permissible" (spark-free) electric locomotives were now required, although these had never before been used west of the Rockies. It took three or four months to get delivery, while mules and horses hauled muck cars on a much retarded schedule. Electric lights and wiring were removed and work went on with just the light from battery lamps on the men's hats. Upon resumption of the work, additional safety measures resulted in further delay, so the 1932 completion date was forgotten and costs were increased over a million dollars.

Hetch Hetchy forces adopted the new safety measures so effectively that a special commendation was received from the state upon completion of the Coast Range Tunnel a couple of years later. "Observance of safety rules is stimulated," according to the 1933 annual report, "by the custom that every week any camp which has not had an industrial accident case enter the hospital during that week is rewarded with a chicken dinner." Those were the days when chicken was the treat, not beef!

Other problems affected work on the Coast Range Tunnel, probably some more serious than the explosion at Mitchell. A totally unexpected problem, however, and one of great magnitude, was an unprecedented period of drought in the San Francisco Bay region. A succession of four dry winters brought about an emergency situation. Storage of water in the City's reservoirs was so depleted that in the fall of 1930 extraordinary measures had to be taken to bring in more water in advance of Hetch Hetchy. The Raker Act had anticipated completion no earlier than 1932, but delays by politicians and the Mitchell explosion had set this date back. Now all the politicians joined in berating O'Shaughnessy.

The East Bay Municipal Utility District was held up as an example. Oakland and her sister cities had withdrawn from Hetch Hetchy in 1924, had found a source of supply in the Mokelumne River east of Stockton, had built a pipeline and brought in water in four years (while "More Money" O'Shaughnessy was still puttering around, went the implication). The critics conveniently failed to point out that the great East Bay system was built in such an expedient manner that its expensive operational methods have been paid for ever since by taxpayers of the area served – an item on the annual tax bill says all that needs to be said! San Franciscans had to have a scapegoat and they jumped all over the Chief.

Three emergency measures were proposed to bring in extra water in a hurry. First, the Sunset District wells were sunk in outer San Francisco and turned out 6 m.g.d. The second, soon implemented, was a supply of 30 m.g.d. from the East Bay Municipal Utility District. This was obtained by constructing a 12½-mile emergency pipeline, mostly right on the surface, with a pumping station taking water at San Lorenzo and delivering it to City pipelines at Newark. Farmers along the route were at first furious, but carefully selected contact men went out and soon pacified most of them. In an unbelievable display of co-ordination, co-operation and speed, this system was put into operation on February 22, 1931, only three and a half months from award of the contract. A million dollars went out of the Hetch Hetchy kitty to pay for it.

As added assurance, anticipating one more dry winter, a third source of additional water was prepared: the Corral Hollow emergency pipeline, to take Hetch Hetchy water out of the Coast Range Tunnel at Thomas Shaft, push it over the hills in a 36-inch pipe down to Alameda West Portal and there put it back into the completed part of the aqueduct. Max Bartell and Jack Best surveyed a route for the 24-mile line, which would have a capacity of 45 m.g.d. Work started in December 1931 under Bartell and ended only six months later; it included a 700-foot tunnel at Seco Summit to save 93 feet elevation. The entire line was done inexpensively, pipe just under the surface and handy for later salvage. And the pipeline was never used! There was a wet winter which built up ample storage in local reservoirs; so the pumps never had to be taken out of their shipping cases. The cost of $951,000 was largely recovered from later sale of the pumps and electrical equipment. The salvaged pipe brought good prices during the wartime shortages a decade later.

Next there was the problem of finances. Here's a quotation from the Department of Engineering's June 1930 annual report: "To conserve funds, pending sale of bonds to finance the work, several headings were shut down for various periods of time." As we have seen, it was a big struggle for an entity the size of San Francisco to tackle such a large project, considering its other problems at the time. The City found it expedient to ask voters to approve only a limited amount of expenditure at a time; so it was necessary to complete a segment and go back, saying, "See, we've got it going. How about some more money to continue?"

Thus bonds amounting to $24,000,000 were voted on May 1, 1928, by a nine-to-one majority, to complete tunnels through the Coast Range and pressure pipes across the San Joaquin Valley. However, in 1929 there was no market for these bonds and that would have been that – perhaps for several years – except that A. P. Giannini had faith in the City and in O'Shaughnessy. His Bank of America picked up a large portion of the bonds as a sort of "public contribution," so the work could continue.

During the first week of February 1932, a crisis was reached on account of the lack of money, and the new Public Utilities Commission "authorized" project employees to contribute 10% of their wages each month for five months "towards sales of bonds needed for payroll to carry on the work." This was done through the Hetch Hetchy Bond Purchase Syndicate, with every person on the job a member. Accountant Willis O'Brien acted as chairman, supervising purchase of bonds from the City Treasurer at par (as stipulated by the City Charter), then selling to the highest bidder at a lower amount. The syndicate paid the difference between par and market value. Whenever possible, contractors and material suppliers took bonds in payment. Thus, again, work continued and some 1500 men were kept employed at the deepest part of the Depression. As far as is known, this was the first time such a plan was undertaken in construction work in this country. The syndicate closed its books after June 30 and paid back to the members 50¼ cents for each dollar contributed. So the "pay cut" averaged out to a net of about 5% for five months – really a minor figure compared to the trend in those years.

In May 1932 a final bond issue of $6,500,000 was proposed to voters in order to complete the aqueduct and bring the first water into San Francisco. It was explained that $2,500,000 had been diverted to emergency pipelines; another million went for extra safety measures and equipment, and almost another to pay the City for Hetch Hetchy power used in construction. To meet demands of the Department of the Interior for the construction of roads and trails within Yosemite National Park, as stipulated in the Raker Act, required $1,500,000, or $250,000 per year for six years. O'Shaughnessy was the leader of a group that went out night after night to speak at various meetings in the promotion of this bond issue. The job was made especially difficult by critical "articles in the noisier newspapers . . . which refer disparagingly of the work in general." This bond issue was also approved by the voters.

The Public Utilities Commission, mentioned above, was brought into being by a new City Charter of 1932, under which there were great changes in the Department of Engineering, in the City Engineer's Office and in all utilities operated by San Francisco (the Municipal Railway, Water Department, Hetch Hetchy, the airport and others). Where previously these utilities had been administered by separate and independent de-

San Francisco's portable Link-Belt aggregate plant, capable of producing 100 tons per hour out of tunnel spoil, built up stockpiles at Alameda Creek, Valle Shaft and in Corral Hollow near Thomas Shaft. It is shown at the latter location in August 1931. *(—City of S.F.)*

This Elwell Parker electric-battery truck was purchased by the City for periodic inspections of Coast Range Tunnel. Carrying 12 passengers, it could travel 72 miles without recharging.

Alameda East Portal in 1930. Narrow-gauge rail tracks emerging from the tunnel connect with the ¾-mile tramway crossing the valley of Alameda Creek. A batch plant has been set up to left in preparation for tunnel concreting work.

Location of Alameda Siphon, connecting the two portions of Coast Range Tunnel where it crossed the small valley of Alameda Creek. Looking east along the narrow-gauge tramway connecting the two portals. At far right sits the Whitcomb diesel locomotive that did "main line" work on this track. *(–Ray Allen photo)*

Track assembly for tunnel concreting is shown at Irvington Portal. The entire layout was moved along on small wheels running on tunnel tram rails. Loaded cars containing batches of dry mix were shoved up from the near end, discharging their contents near the mixer and concrete gun at far end of the rig. *(–City of S.F.)*

The City Hall engineering staff of San Francisco's Bureau of Engineering, 1928. Left to right: Joe Callaghan, designer; Max Bartell, hydraulic engineer; Clyde Healy, Second Assistant City Engineer; M. M. O'Shaughnessy, City Engineer and Chief of Hetch Hetchy; Nelson Eckart, Chief Assistant City Engineer; Henry Ohmer, Chief Designer; Paul Ost, Chief Electrical Engineer. *(–City of S.F.)*

The shaft and camp at Indian Creek, in a country setting south of Livermore, was the final shaft and was five miles west of Valle. Depth was only 304 feet, including the rock pocket below tunnel level. *(–City of S.F.)*

partments and boards, all administrative authority was now centralized in the hands of an appointed commission. One section of the new charter that affected the aqueduct work required that any City expenditure of more than $1000 required bidding and a contract. This opened completion of the Coast Range tunnel to the lowest bidder. Five bids were received, the lowest being that of the "Hetch Hetchy Department," which had been doing the work all along. Its bid was $557,670 lower than the closest competitors, but the losing contractors tried desperately to prevent the award to this bidder. However, the Hetch Hetchy Department got the contract and the City's own engineers, superintendents and tunnel crews kept right on digging westward.

Another effect of the new charter was the removal of Michael M. O'Shaughnessy from the job of City Engineer and head of Hetch Hetchy. His chief supporter, Jimmy Rolph, had left the City to become governor of California. Many, many people have stated that certain sections of the charter were written with just the aim of getting rid of the Chief. This is not at all farfetched, as the Old Man had enemies all over the City. He had thrown newspapermen out of his office, had publicly derided the Board of Supervisors, had even refused to give Hetch Hetchy jobs to relatives of politicians! And he had never tried to make himself look good with a big public relations effort. As James Turner, a later Manager of Hetch Hetchy and Manager of Utilities, pointed out, O'Shaughnessy could have taken that initial $45,000,000 bond issue, stretched the money out on a smaller-capacity project and made it appear to go further, but he refused to be deceptive.

Although they loved the Chief and admired him, several of his associates later stated they believed he was partly responsible for his own downfall, although no one approved of the way it was handled. Lloyd McAfee, who had been Construction Engineer on the Mountain Division

The seventh and final holing-through, completing the then-longest tunnel in the world, took place on January 5, 1934, between Mitchell and Mocho shafts. Seated on the platform, left to right, are Nels Eckart, Les Stocker, Chief O'Shaughnessy, Mayor Angelo Rossi, Lewis Byington (President, P.U.C.), George Filmer (P.U.C. member), T. W. Espey (Water Dept. engineer) and Supervisor J. Emmett Hayden. In front are George W. Pracy (standing, Water Dept. engineer), Edwin Eddy (P.U.C.), Louie Wheeler and Edward J. Cahill (Manager of Utilities). A different picture appeared in the *Examiner* the next day, one that left out Eckart, Stocker and the Chief! *(—City of S.F.)*

and other parts of the project, had become Chief Assistant City Engineer in charge of Hetch Hetchy in 1930. In 1932 the new P.U.C. appointed him Chief Engineer and Manager of Hetch Hetchy to replace O'Shaughnessy. Paul Ost, former Chief Electrical Engineer, was put in charge of power operations. O'Shaughnessy's new title was "Consulting Engineer" with an office in the Water Department Building on Mason Street. "But he wasn't given any work," recalled Jim Turner, who said he considered the Old Man a "Consulting Engineer Without Portfolio." Some told how they were advised not to take their problems to him, so he was rarely consulted. Max Bartell, McAfee and Turner agreed among themselves to keep his interest aroused and they stopped in at least once a week, or accompanied him on inspections of Hetch Hetchy work.

As the giant tunnel neared completion and time for the final holing-through approached, the Chief was most anxious not to miss the event. But in his new capacity he had no way of knowing exactly when this would be, and those in charge were not about to tell him. Buddy Ryan said he got a phone call from the Old Man nearly every evening, to check on the date. Ryan was one of the friends, so when the date was set at last – January 5, 1934 – he had O'Shaughnessy meet him in Livermore. They drove out to the shaft and went down and along the tunnel without anyone knowing. The "big wigs" were already on hand, according to Ryan, hanging around waiting for the great moment. But tunnel boss Pete Peterson held back on the last 12 inches, knowing that the Chief was on his way. Into the light came the motor "with just the Chief and me on it," Ryan recalled, noting that the others were quite surprised. Ryan helped old Mike up onto the platform and Pete pushed the hole through from the other side, leaned through and asked the Chief to shake hands. Buddy Ryan thought the Old Man would burst with emotion. On the ride back O'Shaughnessy said simply, "John, the old gray horse beat 'em."

The final section was holed through 2500 feet below Crane Ridge between the shafts at Mitchell and Mocho. Here had been found the most unfavorable ground. In one place the tunnel had been excavated to 18-foot diameter, solidly supported by 18-inch square timbers, but during a period of 24 hours this squeezed in three feet all around, and in only a few days the tunnel had become so small that it was difficult for a person to crawl through on hands and knees. Gunite could not be placed by the usual method, because the quick squeezing wouldn't allow concrete to set.

A new system of ground support was tried and proved very successful. The tunnel was excavated oversize so that when gunite was placed between two sets of forms there was a space of about one foot into which the ground might swell without putting any pressure on the concrete until it had set. The lining was two and one-half to three feet thick and mixed with a very high cement content; it successfully withstood the enormous pressure, even in places where the tunnel passes directly through the fractured junction of two earthquake fault lines. Periodic inspections in ensuing years have shown the tunnel to be virtually as sound as on the first day it carried water.

Concrete lining on the Coast Range Division was completed about three months after the tunnel was open. A million yards of aggregate was needed for the work, so City forces bought a portable Link-Belt aggregate plant, able to produce 100 tons per hour out of the tunnel spoil, yet capable of being dismantled and moved on flatcars or trailers. It weighed 130 tons, was electrically operated and moved on tracks of seven-foot gauge. Three points had to be serviced by this giant machine so that underground hauling would not exceed seven miles and the deeper shafts could be avoided.

Batch plants were set up at Alameda Creek, at Valle Shaft and in Corral Hollow near Thomas Shaft. Dry aggregate was pumped down newly installed ten-inch pipes at the shafts, falling into two-yard tunnel cars. The combined mixing and pouring plant in the tunnel below operated on the two-foot gauge tracks, as before. Collapsible steel forms were moved ahead and set up using the same rail system. Concreting differed from the mountain tunnels in that here the floor was poured first; sides and overhead followed, using Webb's concrete gun. Some sections of the lining had flexible joints on 10- to 15-foot centers to cope with the known earthquake threat.

Now there followed the usual spate of claims for water loss in grazing areas above the new tunnel, but engineers from the Livermore office had been measuring all springs for several years and hired a consulting geologist to assist them. Every claim was investigated: some were well founded and the owners recompensed; others were unwarranted.

When the accounts were all in, an official audit showed a saving of over a million and a half dollars through having the big tunnel completed by the City's construction division instead of the next highest bidder. Hetch Hetchy working forces had bid a half-million dollars lower and still finished the job with over a million bucks not spent! Appreciation of this great work was shown by Superintendent of Safety C. H. Fry, of the California Industrial Accident Commission, in a letter written in 1933 and quoted in the *Chronicle.* Fry condemned the unjust criticisms that had been made of the progress on this division, and went on to say, "I believe that had these tunnels been started on a contract that the contractor would soon have thrown up his contract and the bonding company would have forfeited the bond rather than attempt to complete [them]."

To make inspections of the tunnel, for which water is periodically shut off, an Elwell Parker electric industrial truck was purchased. It was equipped with seventy Edison A-10 cells for power and made the trip in five to six hours, average running speed being six miles per hour. Under full charge the vehicle was good for 72 miles, carrying 12 passengers and supplies. Engineer Bill Helbush remembered the bumpy ride on some of these inspections. He said the tunnel was always found to be in thoroughly satisfactory structural condition but the inspection was an all-day ordeal.

So pure water from the mountains was ready to flow into San Francisco. For, as we shall see, the failing sources of Spring Valley water had made it imperative that the pipeline across the Bay be completed ahead of schedule.

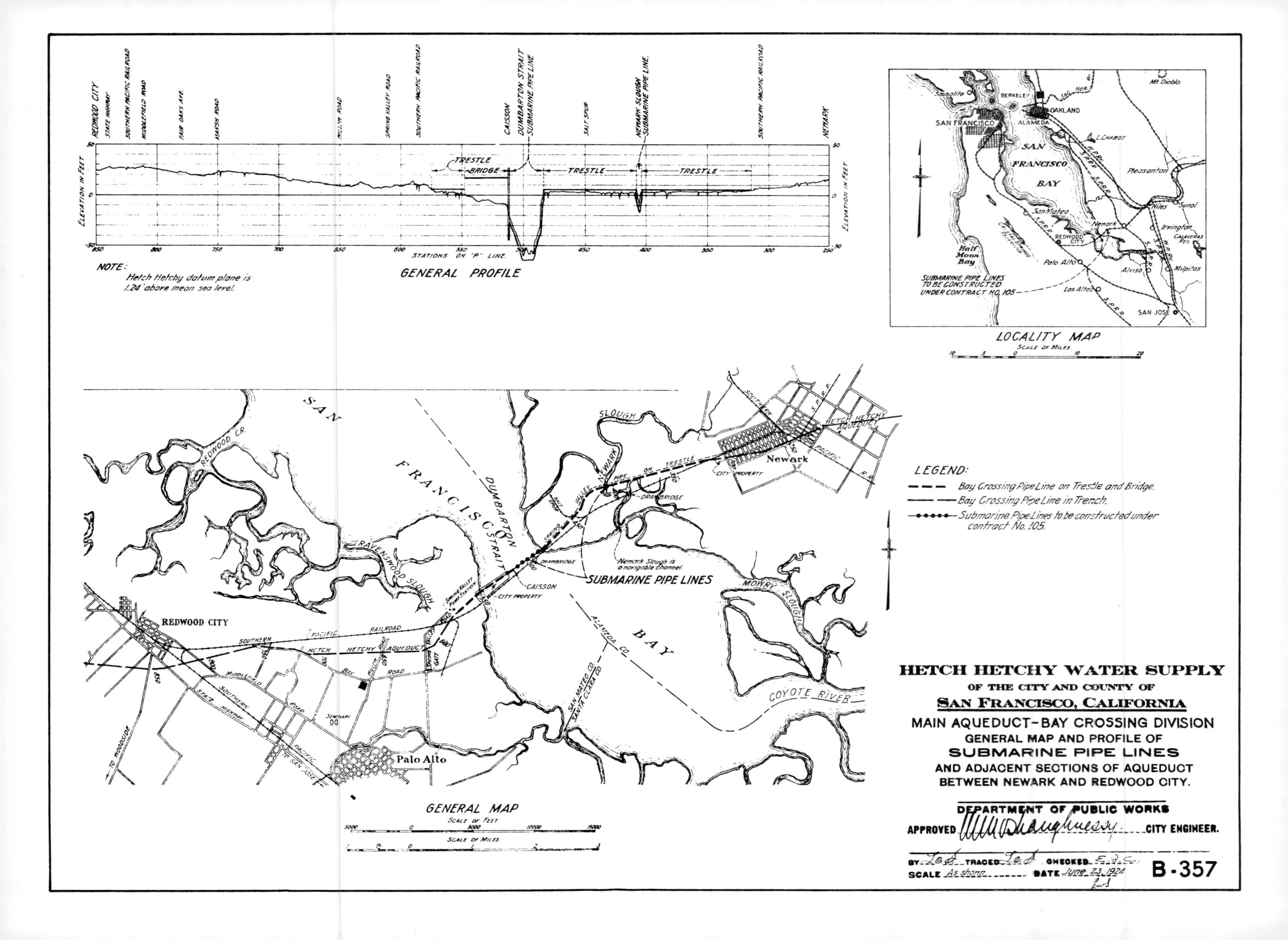

REDWOOD CITY
STATE HIGHWAY
SOUTHERN PACIFIC RAILROAD
MIDDLEFIELD ROAD
FAIR OAKS AVE.
MARSH ROAD
WILLOW ROAD
SPRING VALLEY ROAD
SOUTHERN PACIFIC RAILROAD
CAISSON
DUMBARTON STRAIT
SUBMARINE PIPE LINE
SALT SPUR
NEWARK SLOUGH
SUBMARINE PIPE LINE.
SOUTHERN PACIFIC RAILROAD
NEWARK
TRESTLE
BRIDGE
TRESTLE
TRESTLE
ELEVATION IN FEET
STATIONS ON "P" LINE.
GENERAL PROFILE
NOTE:
Hetch Hetchy datum plane is 1.24' above mean sea level.
Sausalito
BERKELEY
OAKLAND
SAN FRANCISCO
ALAMEDA
L. CHABOT
SAN FRANCISCO BAY
Pleasanton
Mt. Diablo
Niles
Sunol
Irvington
CALAVERAS RES.
San Mateo
REDWOOD CITY
Newark
Half Moon Bay
Palo Alto
Alviso
Milpitas
Los Altos
SAN JOSE
SUBMARINE PIPE LINES TO BE CONSTRUCTED UNDER CONTRACT NO. 105
LOCALITY MAP
SCALE OF MILES
SAN FRANCISCO
DUMBARTON STRAIT
SLOUGH
NEWARK
VALLEY
SPRING
SALT SPUR
PIPE ON TRESTLE
DRAWBRIDGE
CITY PROPERTY
Newark
HETCH HETCHY AQUEDUCT
PACIFIC R.R.
SOUTHERN
S.P.R.R.
Newark Slough is a navigable channel.
SUBMARINE PIPE LINES
CAISSON
CITY PROPERTY
SPRING VALLEY PUMP STATION
RAVENSWOOD SLOUGH
REDWOOD CR.
REDWOOD CITY
SOUTHERN PACIFIC RAILROAD
HETCH HETCHY AQUEDUCT
MIDDLEFIELD ROAD
BAY ROAD
WILLOW ROAD
SPRING VALLEY GATE
STATE HIGHWAY
SOUTHERN PACIFIC
SEMINARY
TO WOODSIDE
TO SAN JOSE
Palo Alto
ALAMEDA CO.
SAN MATEO CO.
SANTA CLARA CO.
BAY
MOWRY SLOUGH
COYOTE RIVER
GENERAL MAP
SCALE OF FEET
SCALE OF MILES
LEGEND:
Bay Crossing Pipe Line on Trestle and Bridge.
Bay Crossing Pipe Line in Trench.
Submarine Pipe Lines to be constructed under contract No. 105.
HETCH HETCHY WATER SUPPLY
OF THE CITY AND COUNTY OF
SAN FRANCISCO, CALIFORNIA
MAIN AQUEDUCT-BAY CROSSING DIVISION
GENERAL MAP AND PROFILE OF
SUBMARINE PIPE LINES
AND ADJACENT SECTIONS OF AQUEDUCT
BETWEEN NEWARK AND REDWOOD CITY.
DEPARTMENT OF PUBLIC WORKS
APPROVED M.M. O'Shaughnessy CITY ENGINEER.
BY L.O.A. TRACED L.O.A. CHECKED E.O.C.
SCALE As shown DATE June 23, 1924
B-357

20

PLENTIFUL WATER AT LAST

As Hetch Hetchy Aqueduct neared San Francisco, it came into the territory of the Spring Valley Water Company, that utility which maintained a strangle hold on the City, had fought and financed the battle against Hetch Hetchy and had actually failed to do an adequate job of supplying water to the people. From 1873 City taxpayers had been able to vote four times on purchase of Spring Valley. But they had rejected all proposals, and the water company meanwhile continued developing sources of supply deemed to be adequate until 1932. It had crossed the Bay into southern Alameda County, drawing 21 m.g.d. from wells in the Livermore Valley and gravel beds on Alameda Creek, where Willis Polk's lovely water temple was erected near Sunol.

Spring Valley started work on Calaveras Dam in 1913, under the watchful eye of Mike O'Shaughnessy, who anticipated that at some future day it would become City property. In that same year he wrote to John Freeman about the questionable and reckless construction methods and designs employed by engineers Mulholland and Hermann, comparing them to the "sad mess" Mulholland made on the Los Angeles aqueduct. The earth- and rock-fill dam, at 220 feet, was the highest such dam in the world when completed. But it had barely started to fill when catastrophe struck: on March 24, 1918, the upstream face sloughed off and slid into the reservoir, taking the outlet tower with it. Mulholland was retired from the project and the dam was finally completed correctly in 1925, with O'Shaughnessy as consultant. (Forty years later the San Francisco Water Department would dedicate the near-by James Turner Dam and San Antonio Reservoir to provide terminal storage adjacent to the main aqueduct.)

An initial five-foot pipeline crossing the Bay was part of the Hetch Hetchy plan, aimed at carrying mountain water to the Peninsula reservoirs of Spring Valley, should that firm be bought out, or all the way into the City if not. However, these reservoirs began to run dangerously short in the early 1920s, and a solution, to be worked out with the water company, was suggested by the City Engineer in 1922. San Francisco would build its transbay pipeline from the Irvington gatehouse to Crystal Springs Reservoir, in advance of finishing other parts of the aqueduct, to carry additional Spring Valley water from Alameda County. So this 22-mile conduit, including pipe above and under the Bay and the 1½-mile Pulgas Tunnel behind Redwood City, was completed in 1926 and leased to the water company. It started delivering 34,000,000 gallons daily from Calaveras at a time when storage was down to a 70-day supply!

In 1928, with Hetch Hetchy water almost on their doorsteps, the voters of San Francisco approved a bond issue to buy out Spring Valley. For $41,000,000 they obtained all water rights, reservoirs and watershed lands, aqueducts and rights of way, and the distribution system in the City. In reaction the *Chronicle* warned that "no better service is expected" because politics would conquer, while Pacific Gas & Electric's *Progress* wrote about "Public Ownership's Back Seat Driver" (meaning politics). The system came under municipal operation on March 2, 1930, with 450 operating employees of Spring Valley transferred to City civil service lists. Placed in charge of San

Crystal Springs Dam and Reservoir, purchased by the City in 1928 from the private water company, is shown in a recent photo with the waters from Hetch Hetchy filling the lakes to overflow. In the background, fog creeps gently over the watershed area from the Pacific Ocean beyond. *(–S.F. P.U.C.)*

Francisco Water Department was Nelson A. Eckart, who had headed construction at Groveland and was later Chief Assistant City Engineer.

Nels Eckart's first big problem was a water supply for several small cities dotting the Peninsula south of San Francisco; their various systems had reached a precarious condition by 1930 and there were thirty or more agencies involved. Engineer Max Bartell had investigated populations and water supplies for the entire area, Daly City to San Jose, from 1928 onward. The various communities and water districts eventually co-operated and have enjoyed Hetch Hetchy water ever since. Public ownership and operation of the San Francisco water supply was a success from the first day, despite the profit-motivated prophets of gloom who had preached disaster. In the first year, after paying interest, redeeming $750,000 in bonds, making additions and starting Upper Alameda Creek Tunnel to add 13 m.g.d. to Calaveras—after all this, there was still a profit of $500,000 put into the City treasury!

When the Coast Range Tunnel was holed through in 1934, junction had already been made with former Spring Valley water supplies nearby in Alameda County, and such supplies were being carried to the Peninsula in the pipeline that had been built a few years early. This Bay Crossing Pipeline was now joined to the western portal of the tunnel at Irvington and for all intents and purposes the Hetch Hetchy Aqueduct was at last complete.

It was an unusual pipeline that crossed lower San Francisco Bay. Mainly of five-foot-diameter

Pulgas Tunnel, through the ridge behind Redwood City, was the final link carrying clear mountain water to the City's terminal storage reservoirs immediately south of San Francisco. This view shows the outfall channel construction, looking westward from the tunnel.

Pulgas Tunnel concrete lining work was done from this portable platform raised above the floor. This tunnel was completed early in 1924, out of the general schedule of work, so that it could carry water temporarily for the private company and lessen the danger of shortages.

The Bay Crossing Pipelines carry Hetch Hetchy and Alameda County waters to the San Francisco Peninsula reservoirs. The view at top, looking east, shows the general layout at west edge of the Bay in 1936. The first line is the covered one in foreground, with then-new second line to left. In left distance is the Dumbarton highway bridge, and to the right, Southern Pacific's railroad trestle and bridge. Above and right are views of the submarine pipeline section being laid from its cradle in 1925. *(–City of S.F.)*

Construction of a steel bridge to carry two pipelines took place in 1924. It spanned that portion of the Bay westward from the underwater section to the shoreline. Crews' camp has been placed on pilings to the left. *(—City of S.F.)*

Hetch Hetchy's Dumbarton Bridge construction, below. Three spans are in place and piers beyond are under construction, August 1924. *(—City of S.F.)*

Bay Crossing Pipeline construction. The first line was laid in 1924-25 and view above shows the 42-inch submarine section being lowered into the waters for placing beneath Newark Slough. To the right is a view of 1936 work on the second Bay pipeline, a concrete anchor being poured at the caisson, where underbay pipes meet those on the steel bridge under tremendous pressure. Below, completed Pipelines Nos. 1 and 2 pass beneath SP railroad track on the eastern side of the Bay, near Newark, after descending from Irvington Portal on the distant hillside. *(—Upper and lower, City of S.F.)*

Hetch Hetchy's Dumbarton Bridge in 1936, showing Pipelines Nos. 1 and 2 in place and carrying water to the City's reservoirs. The smaller line (right) is the 60-inch pipe placed in 1925, while to the left is 76-inch pipe put in service in 1936. Two additional pipelines have been added in recent years, but they take a longer route around the southern end of San Francisco Bay. *(—S.F. P.U.C.)*

steel, it was underground for six miles after leaving Irvington Portal, then mounted a timber trestle over salt marshes toward the Bay. It dipped in a siphon 400 feet long at Newark Slough and in another over a half-mile long at Dumbarton Strait, at which point the bottom of the pipe lies 68½ feet below sea level. At the two navigable crossings mentioned the pipe changed to 42-inch cast iron two inches thick. Entering a massive concrete caisson just west of the strait, the pipe was carried over the rest of the Bay on a steel bridge of 36 spans. Thence it was a half-mile or so to Bay-Pulgas Pumping Station (which was eliminated when full gravity flow started in 1934), a shallow pipeline under Redwood City and up into the hills westward. Elevation 290.5 was reached at the pipe's end, where it joined Pulgas Tunnel.

Over the years other pipes have been built and added as required, Bay Division Pipeline No. 2 being started in 1934, even before Hetch Hetchy water reached the City. It was larger, carrying 65 m.g.d. compared to 43 m.g.d. in the first pipe. No. 2 shared "Hetch Hetchy Bay Bridge" with pipeline No. 1 as planned. Spring Valley's small 16-inch pipe, after 48 years service beneath the Bay waters, was removed when No. 2 was installed; the old pipe was in such excellent condition that it was cleaned, recoated and put into use in City distribution. Two further pipelines have been added between Irvington and Pulgas in more recent years, but Nos. 3 and 4 are each 34 miles in length, passing around the end of the Bay. This reduces the possibility of simultaneous loss from damage by earthquake or other causes. These pipes brought total capacity of Bay Division to over 300 m.g.d.

Hetch Hetchy water first passed through Pulgas Tunnel and entered Crystal Springs on October 28, 1934. Eighty-nine persons had lost their lives in putting together this enduring system. The last to die before completion was Michael Maurice O'Shaughnessy, the old Chief, who had personally

Sunset Reservoir was completed by the S.F. Water Department in the western part of the City in 1959. Since purchase of Spring Valley Water Company in 1928, City operation of the water service has been a tremendous success. *(—S.F. P.U.C.)*

Ceremonies marking the completion of Hetch Hetchy and the first flow of pure mountain water into the City's reservoirs took place at Pulgas Water Temple behind Redwood City, October 28, 1934. The "lasting monument to the energy of the City" was in operation after twenty years of hard work. *(—City of S.F.)*

THE MUNICIPAL RECORD October — October THE MUNICIPAL RECORD

WELCOME TO OUR CITY

WATER FROM HETCH HETCHY

LEWIS F. BYINGTON
President of Utilities Commission

Directions to Hetch Hetchy Celebration

Take Skyline Boulevard to La Canada Road, then a short distance along this road to parking spaces to south at lower Crystal Springs Lake. Or use Route 101, or Bayshore Highway, and cut over from San Mateo to La Canada Road.

O'SHAUGHNESSY DAM

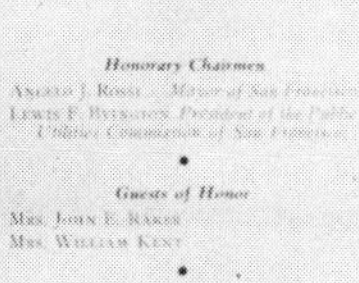

Honorary Chairmen

ANGELO J. ROSSI ... *Mayor of San Francisco*
LEWIS F. BYINGTON, *President of the Public Utilities Commission of San Francisco*

Guests of Honor

MRS. JOHN E. BAKER
MRS. WILLIAM KENT

SUPERVISOR FRANCK R. HAVENNER
Chairman

THE LATE M. M. O'SHAUGHNESSY

JACKSON T. CARLE
Director, Bureau of Public Relations, Public Utilities

EDWARD G. CAHILL
Manager of Public Utilities

PROGRAM

Introduction of Hetch Hetchy Builders
Address .... LEWIS F. BYINGTON, *President, Public Utilities Commission*
Release of Hetch Hetchy water at Dumbarton
Address .... HAROLD L. ICKES, *Secretary of the Interior*
Arrival of Hetch Hetchy water
Concert .... SAN FRANCISCO MUNICIPAL BAND—PHIL SAPIRO, *Director*
Welcome to Hetch Hetchy .... MAYOR ROSSI
Tribute to M. M. O'Shaughnessy .... SUPERVISOR JESSE COLMAN, *Chairman, Public Utilities Commission, San Francisco Board of Supervisors*
Solo .... JACK HOWELL, *Baritone*
Labor's Part in Hetch Hetchy .... SENATOR DANIEL C. MURPHY, *Member, Public Utilities Commission*

N. A. ECKART
General Manager and Chief Engineer

SUPERVISOR FRANCK HAVENNER

The Following Professional and Business Men and Firms Have Helped to Make This Hetch Hetchy Water Edition a Success

SPRING VALLEY COMPANY, LTD., *S. P. Eastman*
MOUNTAIN COPPER COMPANY
HAMMOND LUMBER COMPANY
GEORGE BISTANY, *Director of Zoo*
HARTFORD ACCIDENT & INDEMNITY COMPANY
CALIFORNIA FILTER SERVICE COMPANY
W. F. CORDES, *Capitalist*
THE DE VILBISS COMPANY
JENISON MACHINERY COMPANY
FEDERAL GARAGE
G. GHIRADELLI COMPANY
POST-TAYLOR GARAGE
CHAS. R. McCORMICK LUMBER COMPANY
GEORGE A. ELLIOTT, *Engineer*
ROBERT W. HUNT COMPANY, *F. M. Randlett, Engineer*
SIERRA EQUIPMENT COMPANY, *Frank L. Du Frane*
PACIFIC COAST AGGREGATES, INC.
ELMER E. ROBINSON, *Attorney*
THOS. M. FOLEY, *Attorney*
THOS. J. RIORDAN, *Attorney*
CHAUNCEY TRAMUTOLO, *Attorney*
W. C. G. McDONNELL, *Attorney*
E. J. MAYFIELD—PACIFIC BONE & FERTILIZER
JOHN P. DEGNEN, M.D.

With the completion of Hetch Hetchy, danger of water shortages in San Francisco and the Peninsula cities was eliminated for at least a century. Here Crystal Springs Dam spillway carries the overflow from filled reservoirs after decades of uncertainty. *(—S.F. P.U.C.)*

A recent addition to the Hetch Hetchy system is San Antonio Reservoir and James Turner Dam, providing additional terminal storage in Alameda County. *(—1967 photo from S.F. P.U.C.)*

seen the project through twenty years of building and was scheduled to be the honored guest on the great day. Early on the morning of Friday, October 12, the Chief complained of a pain over his heart and died before a doctor arrived – only 16 days before his great dream was to become a reality. He had reached the age of 70.

The *Call-Bulletin* headlined "M. M. O'Shaughnessy, Job Ended, Dies" and told in affectionate terms of his devotion to the City and to Hetch Hetchy. His coworkers referred to him as "the old man of the mountains," said the paper, but he was "M.M." to everyone else in San Francisco. His successor, Lloyd McAfee, wrote in *The Municipal Record,* "The great genius and mastermind who alone was responsible for the prosecution of the work, who fought for it during the whole period of his employment by the city as Chief Engineer and Consulting Engineer, has passed away. . . . Let every San Franciscan honor him and revere him as we do, who worked so long with him." Some of the very politicians who had been foremost in shoving him brutally to the side two years before were now quoted by the press as eulogizing him to the heavens.

Seldom has the passing of a city official brought such an outpouring of sorrow, and never such a listing of accomplishments. The stately *Engineering News-Record,* published in New York, commented on the tragedy of his dying just before completion of his greatest work. The journal mentioned many engineering successes he'd had before O'Shaughnessy came to the San Francisco job and pointed out that this man "was particularly well known for his disregard for politicians and political methods in the carrying out of engineering work."

Shortly before his death M. M. O'Shaughnessy published a book telling the story of planning and building Hetch Hetchy. Anticipating arrival of the mountain water a few months hence, he wrote, "The final result was eminently satisfactory to the people and the taxpayers of San Francisco, and they are the proprietors for whom I have been working and who have to pay the bill."

They gathered at Crystal Springs Lake in San Mateo County on October 28, 1934, to hear the Municipal Band and speeches by Mayor Angelo Rossi, Supervisor Franck Havenner and Harold L. Ickes. A tribute to O'Shaughnessy was printed in the official program and another was spoken by Supervisor Jesse Colman. A nation-wide radio broadcast carried the exercises, which were held before the specially constructed Pulgas Water Temple. There was wild cheering from thousands of spectators as the first water flowed over the weir and down toward Crystal Springs – one of the epochs in San Francisco's history.

During the final year of work a most important letter of commendation had come from J. Waldo Smith, who for many years served as Chief Engineer and Consulting Engineer of the New York Board of Water Supply. He expressed "unqualified admiration for the conception and execution

Calaveras Dam on Alameda Creek, southern Alameda County. Started by Spring Valley Water Company in 1913, the dam suffered a partial failure in 1918 and was completed under the direction of Chief O'Shaughnessy in 1925. It was taken over by the City when the private company was bought out in 1928. The view is looking westward in 1936. *(—City of S.F.)*

. . . of the aqueduct." Further, Smith continued, "When the project is completed, it will stand as one of San Francisco's greatest assets, and as a lasting monument to the energy of the City and its citizens, and their faith in the future." Perhaps more than anything else, this man's words expressed the realization of what San Francisco had accomplished.

The greatest water project ever undertaken by a municipality was now in operation, 20 years and $100,000,000 after the start of construction way back in 1914. The authoritative *Water Works and Sewerage* magazine termed it "one of the world's most important engineering projects." And this was just the end of the beginning, as it were. Already a second pipe was in progress across the Bay, and already plans were almost completed for making massive O'Shaughnessy Dam taller.

The sale of construction equipment and supplies no longer needed had gone on almost continuously over the years as various segments of the Hetch Hetchy Project were completed. For instance, the railroad disposed of the two Heisler locomotives in 1923, and other equipment passed out of the picture when buyers could be found. As the last tunnels were completed in 1934, the trains and rails, mucking machines, drills, motors and compressors became redundant. They were sold through the office of the purchaser of supplies, in accordance with charter provisions.

Despite newspaper predictions and a distribution system complicated by the many hills that make San Francisco so attractive, water rates actually went down and stayed down after the people took over and "socialized" the system. And in spite of enormous population growth in suburban areas served by San Francisco Water, the number of employees in this department increased very slightly over Spring Valley personnel. In spite of these economies, the Water Department in the City is one of very few major water suppliers in the United States which is completely supported by revenues from customers, with no tax subsidies, hidden or otherwise. "San Francisco is the last big American city which still delivers into its domestic distribution system water which is pure enough to require no chemical treatment for health reasons," according to a recent magazine. Actually, there is mild chlorination just for health insurance and the water has been fluoridated since 1951. But the pioneer leaders of this city, who searched for and fought for and finally delivered the Hetch Hetchy water system, left a legacy which any municipality can envy.

Hetch Hetchy Railroad continued train operation through the 1930s and '40s by the use of three Plymouth gasoline locomotives. Largest, shown above, was a twenty-ton model bought new in 1931. Bearing the number 2 at first, it became No. 1171 in a later renumbering into the City's equipment list. It was photographed at Moccasin just before abandonment in 1949. Shown to right is No. 1170, an eight-tonner originally numbered 3 when bought secondhand in 1934. Below is No. 1176, another eight-ton Plymouth, bought new in 1928 and originally numbered 1 (for a short time it bore the number 10). The small locos were photographed at Moccasin in 1948, by which time they were operating relatively few miles per year. All three were painted orange and carried the emblem lettered "SF – Hetch Hetchy Power Division."

21

HETCH HETCHY RAILROAD LIVES ON

When Foothill Division work was completed and headquarters at Hetch Hetchy Junction dismantled, in 1930, the center for mountain operations was set up at Moccasin. Here the City owned plenty of land for shops, garages, warehouses and homes, and the now-private railroad had plenty of room for its remaining equipment and operating headquarters. The only surviving steam engine, No. 5, was stored at the junction, awaiting sale. A light eight-ton Plymouth gasoline locomotive, purchased for work at the junction shaft, was found able to move single-car freight trains on the main line. (A 1931 contract with a road builder states that HHRR agrees to haul the contractor's heavy equipment to Mather, using the eight-ton engine and one flatcar. The load limit would be 15 tons; a round trip would take two to three days; cost would be $120 for the round trip; no foreign cars; no passengers.) A larger Plymouth, twenty-ton No. 2 (later No. 1171) was put in service late in 1931 and a second eight-tonner in 1934. But why would the City of San Francisco invest in motive power, keep a few track trucks and railcars running or even hold on to a 59-mile industrial railroad several years after construction work was completed?

The railroad was needed for maintenance of power lines, for supplying and maintaining camps and other facilities, especially in winter. There was always a busy camp at Early Intake: the small powerhouse still produced; it would be needed for construction again when the dam was raised. Operators and their families lived here, also crews to maintain the aqueduct bringing water to the powerhouse from Cherry River. Starting service in 1918, this system had originally consisted of a mile of concrete-lined canal, a mile and a half of wooden flume and five tunnels aggregating another mile. Early in 1922 the side of a hill slid away, leaving a 500-foot gap in the canal. The wooden flume portion, which included the powerhouse forebay, needed constant maintenance.

Finally the flume was replaced by pipe in 1931. This job required extensive use of the Hetch Hetchy Railroad and construction of a new narrow-gauge tramway. The pipe was hauled in two-car trains by Plymouth No. 2 to Intake Siding at milepost 51, a round trip taking 16 hours. Pipe was transferred by crane to the car on the Intake incline and lowered to the bank of the Tuolumne River. There it was picked up again, by aerial cableway, and taken across to a pipe bench, on which a two-foot-gauge rail line was used to haul it to its place in the trench. Larger pipe transported water from the last of the lined tunnels to the penstock, while smaller sections continued along the bench to Intake Diversion Dam. Thus surplus runoff went into the aqueduct for use at Moccasin Powerhouse and on to the City. By 1935 the concrete ditch portion of this small aqueduct was replaced by tunnel, finally eliminating a constant source of annoyance – cattle and deer getting caught and swept toward the penstock.

A new warehouse was built at Early Intake Siding in 1934 just to take care of supplies delivered by rail. At the same time that fearsome 4½-mile road from here down to Early Intake was improved to the point that "now two cars can pass." Other roads serving project camps in the early years of the Depression were improved on a continuing basis by State Emergency Relief Act (SERA) labor.

No. 19, the former ambulance, and truck No. 21, a former passenger rail-car, were caught in a snowstorm near the top of Red Hill in the winter of 1931-32. The vehicles were placed beside the track on some timbers and retrieved a month later when the snows had melted. *(—Moccasin Powerhouse collection)*

In the early 1930s rail traffic was light on the Hetch Hetchy, though the line had to be kept open in winter-time to provide maintenance and communications. Speeder No. 30, shown left, west of Mather, took a couple of employees' kids along for a thrilling ride over the soft snow.

Above is the Intake Tramway car ready to make its descent into the canyon after a fresh snowfall. Laden with supplies, the car was used as a battering ram against piled snow. *(—Both, Moccasin Powerhouse collection)*

Winter maintenance "train," when the snow wasn't deep enough for a rotary, consisted of track bus No. 24 and a pusher. After the end of passenger service No. 24 served for many years as the winter line car. Its rear-mounted engine was removed and one of the eight-ton Plymouths was attached at the back to shove it through the snow. Viewed a half-mile east of Groveland, January 1930. *(—City of S.F.)*

Late in 1931 Lower Cherry Aqueduct, which supplied water to the small powerhouse at Early Intake, had sections of steel pipe installed in place of deteriorating wooden flume. The pipe arrived by Hetch Hetchy Railroad, was sent down the cable incline, swung across the Tuolumne River and then distributed along the bench on a two-foot-gauge tramway using this unusual gasoline locomotive. *(–Moccasin Powerhouse collection)*

The job didn't come to an end when O'Shaughnessy Dam was completed and water sent rushing down tunnels and pipes. There were improvements to be made, constant maintenance and repair, and always the long-range plans for making the big dam 85½ feet higher. It was found that a railroad was the only practical means of getting back into those mountain areas in wintertime. Some of the roads and highways were still not surfaced into the 1930s and '40s; they were snowbound for as many as three or four months. On the other hand, the railroad provided a smooth, solid "road" that could be cleared by snowplows and used all year. But the line was kept in a sort of minimal condition, suitable only for operation of its gasoline-propelled equipment.

The railroad was under the management of the Power Operating Division at Moccasin, because until the mid-1930s it was used almost entirely in connection with their work. However, in the first four years after discontinuance as a common carrier in 1925, very little maintenance work was done. In June 1929 Thornton Easler, Superintendent at Moccasin, wrote to City Hall. He described terrible conditions observed on some portions of the railroad, emphasized that it had to be kept in operating condition and laid out plans for holding it together. In some places it was twenty to thirty feet between dependable crossties! Several cattle guards had been broken when the steam crane was taken out to the junction in 1929. A trestle over the highway at Big Oak Flat had rotting sills. There was absolutely no question that the road had to be maintained in condition for operation. To keep the power system going required telephone communication between Moccasin, Early Intake, O'Shaughnessy Dam and Lake Eleanor. The phone line followed the tracks, as did the 22 kv power line from Intake, so the railroad was the only means of reaching them for inspection and repair.

In those days there were pieces of equipment in Moccasin Powerhouse which could not be taken out for repairs, nor could they be replaced, without the rail line – not with highway bridges limited to five-ton loads! Also, City officials in the mountain areas claimed they made big savings in hauling such items as lumber, large pipe, crushed rock, cement, etc., by using their existing railroad. Paul Ost, Chief Electrical Engineer in the City, talked with people at Pacific Gas & Electric about keeping roads open to their hydro plants in winter, and he was told it was exceedingly difficult and sometimes impossible. They considered that Hetch Hetchy was in an enviable position with rail tracks handy to most sites and equipment to keep it open at minimum expense. Ost's report to the utilities manager pointed out one simple fact: "The revenue from our power houses is over $6,000 a day so that either the loss of water or the failure of any of the control devices for want of proper communication and transportation would quickly exceed the entire cost of railway maintenance." (It was averaging under $20,000 per year with all gasoline equipment.)

In the early '30s deteriorated bridges and trestles were repaired, every fourth or fifth tie replaced and drainage ditches, culverts and cattle guards restored. The shop crew at Moccasin designed sliding windows to replace the leather side curtains of famous old track ambulance No. 19 and it became a mild-weather linecar. De luxe

Lower Cherry Aqueduct, carrying water from the Cherry River, was often damaged by slides. Portions of both the wooden and the concrete flumes were swept away, sometimes in only a few minutes. Over the years the line was improved and strengthened by using steel pipe on trestlework and by putting more of it in tunnels. These views show the disastrous slide of 1922, which put the powerhouse out of service during a vital construction period. *(—Right and lower, Doug Mirk photos from Vince Martini)*

Caboose No. 16 was shoved off the end of Intake siding one day about 1930, with the near-fatal results shown to left. The car body, however, was salvaged, taken back to Moccasin shops and used to build a gas-powered rotary snowplow. Below is the result: rotary snowplow No. 16 working its way up Big Creek Canyon in winter 1932-33. It really threw the snow out away from the track, which was the answer sought for so many years. *(—Upper, Dud Snider; lower, Moccasin Powerhouse collection)*

track bus No. 24 had its rear engine removed and a wedge plow fitted to the front. With one of the light Plymouths shoving at the rear, it became the foul-weather trouble shooter car and snowplow. Buses Nos. 20 and 23 were stored (23's last run was in 1931) and No. 21 had been rebuilt into a rail freight truck, operated for years by Walt Mitchell and Ernie Beck hauling supplies out of Hetch Hetchy Junction and Moccasin, even coal to the hospital at Groveland.

Pierce-Arrow No. 7 was the last of the older rail trucks to survive. Fitted with a tank, it carried bulk oil for heating plants at Moccasin and other locations, but not for long. Merle Rodgers said you could always tell when the 7-spot was coming – it smoked like a steam engine! Del Gilliam jokingly recalled that it seemed to burn as much oil as it hauled. He was driver of No. 7 on what turned out to be her last run in 1929, when she caught fire opposite Jacksonville and was hauled to the scrap heap.

One economy in the use of railroad equipment by Hetch Hetchy, obviously, stemmed from the fact that its vehicles could haul so much more up the relatively light grades of the railroad than could road trucks on steep, twisting Priest Grade and other notorious places along Big Oak Flat Road. Fortunately for those interested in transportation equipment and history, there were few or no high-pressure salesmen tearing around the countryside in those days, bribing equipment managers to "modernize" their operations.

Hetch Hetchy train sheets of the early '30s, duly signed by the powerhouse operator on duty, acting as dispatcher, show virtually daily activity on the railroad. For instance, "motor 18," a heavy speeder, was dispatched from Groveland to milepost 33 to put in a day on track maintenance, then phoned in and was cleared for the return. The same date shows "Extra 21 West," the supply truck, from Moccasin to Hetch Hetchy Junction, returning as "Extra 21 East."

When there were actual cars of freight to move, No. 2 did the work. The 20-ton locomotive often hauled two-car, 80-ton trains up the grades. Dud Snider, who was in charge of shops and transportation, said he once hauled 160 tons of equipment up the big hill with the 2-spot, but she was in low gear all the way. She averaged about one mile per gallon of gas. Hank Femons saw this same rugged engine drag a train of eight empties back from Mather, though they had trouble on the curving climb west of South Fork. "Sometimes there would be no locomotive call for two or three weeks," Walt Mitchell said, then they'd be running on a job for maybe 15 days in a row (the men all held regular jobs at Moccasin when not operating trains). It would be down in the morning to the junction to pick up carloads of, say, pipe, then back and up the big hill and perhaps reach Priest at the end of an eight-hour shift. The train would be left there and its crew picked up by auto for the return to Moccasin. Next day the trip would be completed and another started on the third morning. Such operation would have been impossible, of course, with steam power. The men who worked with the gasoline railroad rigs were exceedingly fond of their equipment; their pride in its operation showed in interviews as much as forty years later.

Curtis "Red" Fent went to work as a mechanic at Moccasin in 1931. He said it was a wonderful place to work, the "best of all city employment," out from under the feet of politicians and newspapers! And he loved the equipment, those wonderful White Motor Co. road trucks with roller bearings in the engines, the track trucks, the rotary snowplow they built at Moccasin, and the great Plymouth locomotives. The Plymouths were renumbered into the City equipment series in the early '30s; so the twenty-tonner became No. 1171, and the small ones Nos. 1170 and 1176. The latter, Fent said, had hand brakes; when let out of gear, they rolled downhill so fast it would make your "hair stand on end." Men running them learned just how fast to hit the turns without going off the track.

Winter was the trying season for Hetch Hetchy Railroad crews, even in the off years between dam-building jobs. During big-time operations of the early 1920s, the railroad used heavy wedge snowplow No. 101, but this required steam engines to push it at speed, so it was laid aside at Groveland when steam operation ceased. For line maintenance in winter thereafter, various experiments were tried, the best being the large wedge plow on former track bus No. 24, shoved by a Plymouth. There were many amusing incidents. Les Phelan remembered the night truck No. 21 was backing from Intake, followed by No. 19, both trying to get back to South Fork. Near the top of Red Hill No. 21 came off the rails in the snow and couldn't be put back. She and the railcar were stuck there with the snowplow working up above. Both motors were wrestled off the line and placed on some timbers – at night, in a snowstorm! – then 16 men climbed onto the small Plymouth at Red Hill crossing for the ride to Groveland. The clutch was burned out at the head of the grade west of South Fork, but they made it back. Nos. 19 and 21 sat there on the hillside for over a month. But, like all the wonderful Whites they were never hard to start – just one or two cranks, even after sitting all winter!

A rotary snowplow licked the winter problem on the Hetchy. Originally planned to be mounted on front of No. 1171, it materialized as a rotary like none seen on any other railroad, designed and built in Moccasin shops when an accident damaged caboose No. 16 in 1930. Construction supervisor Gene Meyer and his carpenters did the necessary body work; transportation boss Dud Snider supervised installation of the Snow King highway plow, a motor and a drive mechanism. The 50 hp motor that came with it wasn't powerful enough, so eventually the vehicle was hauled back to Moccasin and got a real power plant: a six-cylinder marine Speedway all-aluminum engine

The rotary snowplow, then numbered 16, is working its way at two miles per hour through fresh snow at Big Oak Flat, in winter 1932-33. The rotary was originally equipped with a small motor, inadequate for railroad service. During the following summer a heavy marine engine was fitted and the problem was solved. *(—Moccasin Powerhouse collection)*

that had reportedly been removed from a vessel at Lake Tahoe. With transmission the engine was eight feet long, had three ignition systems and produced 350 hp which drove the shaft of two snow-throwing wheels by means of an eight-inch-wide "silent chain" in oil bath. A rack to one side held the 50-gallon gas tank, feeding engine fuel by gravity. On the roof above was mounted a big radiator; the extra-long fan belt passed through a chute in the engine-room ceiling. An exhaust pipe also went through the ceiling.

At the rear a slightly larger room made up the living quarters, because the rotary often worked out on the road for several days after big snowstorms. The room had a stove, table, sink and several stools. Gilbert Leach's job included being cook on the rotary and he had to feed eight or nine men on the usual turn: plow operator, signalman (to watch for obstructions ahead), locomotive engineer, a couple of linemen and three or four laborers. At 1200 rpm they were able to plow dry snow up to five feet deep, or three feet of wet, frozen snow. On the average they cleared a mile an hour in fairly heavy snow three to four feet deep. Small "wing" snow deflectors in front of the wheels could be raised and lowered from inside; with turnbuckles and cables the entire front of the vehicle could be raised from two to six inches for traveling without plowing. The lookout man usually stood out in front between the chutes, ready to signal "stop" with a bellcord to the locomotive. It was often necessary to back up when bucking snow in cuts, in order to get clean rails, because the engine would soon be up on ice if going too slow.

Usually a plowing crew made it to Mather by 9:00 or 10:00 at night, according to Red Fent, and they all slept in one of the empty cabins with a roaring fire going. Next morning it would be necessary to build a fire on the roadbed underneath the locomotive crankcase and pour boiling water into the radiator. Then they'd put a rope on the flywheel and the men would run out, almost like a tug-o-war! It was nearly always too cold for

On its first run rotary snowplow No. 16 was tried out after a wedge plow had gone over the line. The view above, from top of plow, shows the V-plow path ahead. That compacted snow beside the tracks could now be lifted and thrown to the side, eliminating the need for closing down or shoveling for days. The views to right and below show the rotary after a heavier engine had been installed and radiator placed on the roof. In these scenes of about 1934, the plow works its way through the forests and triumphantly reaches the Mather station, after which the yard tracks were cleared and the tired crew returned to Groveland. *(—Above, Moccasin Powerhouse collection; right and below, City of S.F.)*

the battery to turn the engine over. One other detail of the Mather layover was dreaded: everyone had to pitch in and dig out five or six feet of snow on one leg of the wye so the engine and plow could be backed around and headed toward home. Sometimes the plow would be five days making a round trip from Groveland to Mather and back, but vital communications were kept open and supplies delivered.

Del Gilliam of Groveland often told about how cold it got on the locomotive when he was carefully shoving the rotary through snowdrifts. Once when Les Phelan came back to relieve him, his glove remained frozen to the throttle! He remembered one particularly bad trip, when the track was icy and they did a lot of slipping. On the way back the Plymouth ran out of gas at Buck Meadows. The temperature was six above zero as the crew walked over to Big Oak Lodge and carried gasoline back to the locomotive. Then there was the time around 1932 when a gang of men under Red Wanderer was at work in late autumn clearing the lake area at Mather. They were caught by an unexpected snowstorm, unable to drive out, so Wanderer walked down the rail line eight miles to Intake Siding and phoned for help. It took 100 hours for relief to get there with the rotary. Two Ford dump trucks and Red's car were shoved up onto a flatcar. His thirty laborers were crammed into the rotary, into the truck cabs, into the locomotive – anywhere they could fit – and all haste was made back down to the lower levels at Moccasin, "where they got thawed out."

An unusually severe winter period started in January 1933. In mid-month there was no snow on the ground; on the 30th it measured 102 inches! Roofs were crushed and phone lines knocked out. Even the paved Big Oak Flat Road was closed east of Groveland until February 15, then opened only as far as South Fork. Beyond that there was only the Hetch Hetchy Railroad for several weeks, and the then-new rotary made it possible to keep the main line open to Mather except for only a few days.

Since the railroad was the only means of transportation sometimes for weeks, it was the policy to pick up and carry anyone who needed a ride and who was able to "take the bumps" on the snowplow train. Dud Snider remembered one time picking up a trapper beyond Intake, several stranded phone company linemen on another trip and often Hetch Hetchy employees suddenly unable to get back to their stations. People who lived at Smith and at Jones Station were occasional hitchhikers on the plow train. The railroad remained a vital institution to people living in its backwoods area, at least in the worst winter months. It provided a never-failing lifeline to the outer world.

But there were no more summer travelers on the railroad, no more tourists or campers. People now drove their cars to Yosemite, to the Oakland and Berkeley summer camps and to the City's new Camp Mather. They could make it from the City in eight hours by using the new San Mateo-Hayward Bridge. They drove across the broiler-hot valley, running-board luggage racks crammed full. The "Thirsty? Just Whistle" signs and Burma Shave sequences kept interest alive (a particular favorite east of Oakdale went Don't Stick / Your Elbow / Out So Far / It Might Go Home / In Another Car / Burma Shave). The Mather summer camp was improved each year as additional cabins were made available, a sewer system installed, the millpond converted to a swimming hole.

So it went in the quiet decade from 1925 to about 1935. Camp Mather boomed in summer; the railroad fought the elements in winter; Moccasin was busy as headquarters all year around. The rail line was kept in a gradually declining state of repair, while the inclined tramways at Intake and Moccasin were well kept up, the former as a needed supply line and the latter to facilitate inspection and repair of the vital hillside penstock line. There were lawsuits regarding water rights along the lower Tuolumne and on the San Joaquin River – not unanticipated, but all meaning extra work. Although Hetch Hetchy Project construction had left the mountain areas and reached the City, there were always projects under way and men at work "up in the country."

In order to get full benefit from power generation, it was necessary to raise O'Shaughnessy Dam at Hetch Hetchy – that is, make it 85 feet higher. The foundation and other work for this addition had been built into the original dam in the early 1920s; now from 1935 to 1938 dam builders were again working at Hetch Hetchy. Above, cutting a fissure in the north cliff face to fasten the dam's higher ends. *(–City of S.F.)*

The Emergency Relief Act of Depression years sought any type of municipal or state improvement work for thousands of unemployed. Several hundred were put to work rehabilitating Hetch Hetchy Railroad in preparation for heavy train operations during the dam raising work. Old HHRR boxcars were converted into bunkhouses for these men and placed along the line. Here is one of them at Moccasin in 1937, after the work was finished. *(–Bill Pennington photo)*

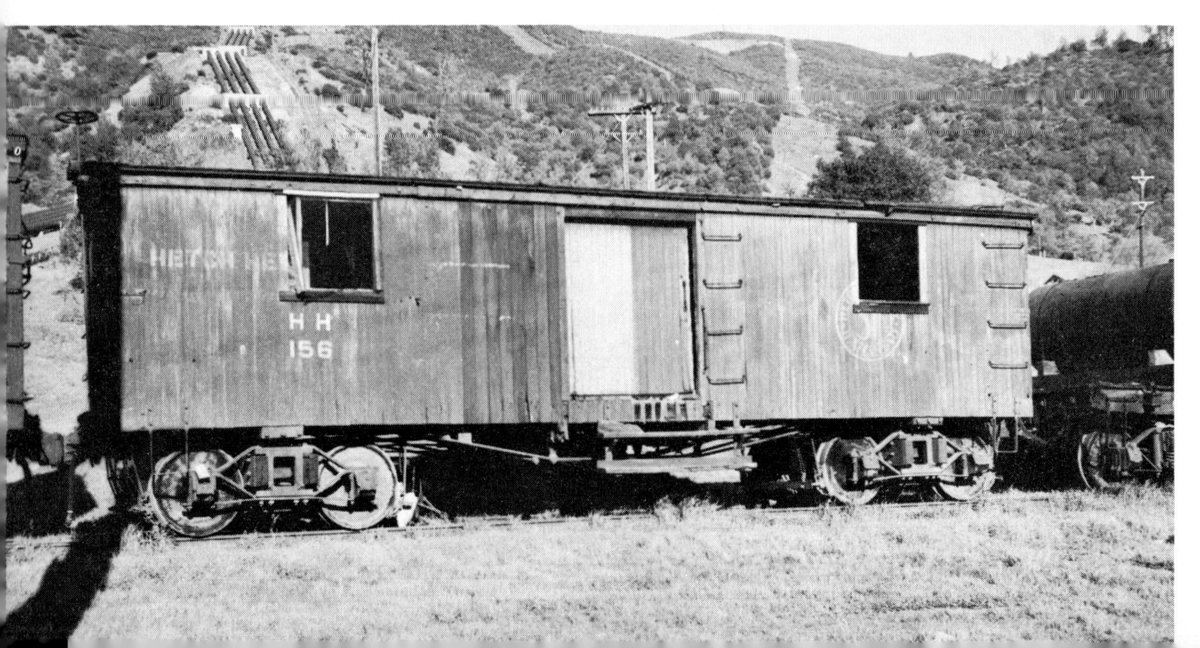

22

SIERRA RAILWAY HELPS TO RAISE THE DAM

Completed in 1923, O'Shaughnessy Dam at Hetch Hetchy had been designed and built with the idea of eventually adding additional height to it and thereby increasing the reservoir capacity by more than 75%. This was desired in order to keep Moccasin Powerhouse running at full capacity the year round. Almost immediately upon the inauguration of Franklin Roosevelt as President, the improvement work became economically feasible through the federal NRA (NIRA) program. The offer of a grant of 30% of the cost of labor and material so changed the financial aspect as to assure that the project would be self-liquidating. Plans and drawings were hurried; San Francisco voters in November 1933 authorized $3,500,000 in bonds; everything was ready for bidding in mid-1934. Bids were received the following January and a contract awarded to the Transbay Construction Co. on a low bid of $3,219,965. One stipulation was that no San Francisco labor be hired until all available men in Tuolumne County were put to work.

Among other preparations, the Hetch Hetchy Railroad had to be readied for steam operation and the hauling of heavy cement trains once again. This had, of course, been anticipated and the management had fought for years to keep the line in shape so it could be rehabilitated for big jobs planned for the future. In 1931 there had been a suggestion to accommodate steam trains of the Sierra Railway, hauling to a dam project northeast of Groveland. On an inspection trip City officials and William H. Newell for the Sierra found that the road was in very poor condition and would need extensive work. That job never materialized, but this inspection and poor report by the Sierra undoubtedly helped Hetch Hetchy operating people present their case for better maintenance. A little later help came from an unexpected source.

It was the time of the Great Depression; thousands of hungry, jobless men roamed the country looking for any kind of work at any rate of pay. On the last day of 1931 the State of California opened its first camp to employ some of these men in environmental improvement projects. This first camp was 15 miles east of Groveland, near South Fork on the Hetch Hetchy. Including food, lodging and token wages, the cost to the state was 28.1 cents per hour.

In December 1933, when the Hetch Hetchy Railroad knew it would have to start full-scale rebuilding, the State Emergency Relief Act (SERA) was asking if the City could put 500 to 600 men to work in "maintenance of municipal property" in the mountains of Tuolumne County. The City could indeed! By the end of March there were 696 men installed in seven camps: Hetch Hetchy Junction, Moccasin, Groveland, South Fork, Early Intake, Mather and a Work Train Camp. At Moccasin, Groveland and South Fork the bunkhouses were old railroad cars converted into living quarters. The City supplied the camps, cots, blankets, water and electricity. Federal funds, administered by the state, supplied room and board for the men, plus cash wages of $5 to $13 per month for 30 hours of work a week. (When WPA was established in September 1935 it took over these SERA programs around the country.)

Locomotive fuel-oil tanks were rehabilitated at Groveland, Hetch Hetchy Junction and Mather; water tanks were provided at the junction, at Cavagnero, Groveland, South Fork and Mather.

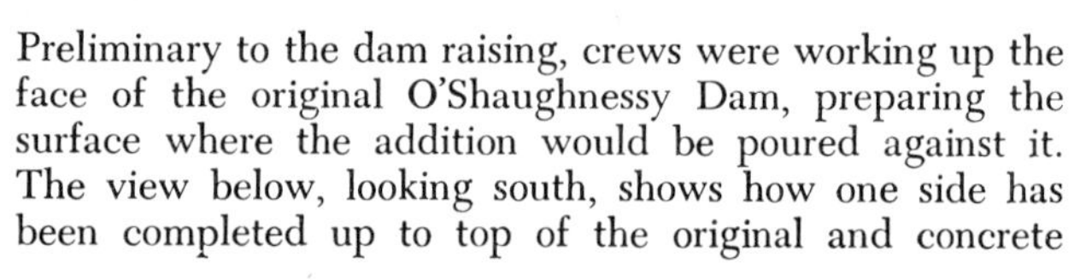

Preliminary to the dam raising, crews were working up the face of the original O'Shaughnessy Dam, preparing the surface where the addition would be poured against it. The view below, looking south, shows how one side has been completed up to top of the original and concrete pouring is in progress on the portion in foreground. In this view, December 1936, top left is the quarry; top right, the aggregates plant. *(—Above, Bill Firmstone; below, City of S.F.)*

This short coach was bought from Sierra Railway in the mid-'30s, when HHRR was preparing for the renewal of heavy train operations. Photographed at Moccasin in 1946 and now the property of the Railway & Locomotive Historical Society, the car is again running on the Sierra and has appeared in scores of films and TV shows.

On December 4, 1937, Al Rose was a passenger on the Sierra freight train at right over Hetch Hetchy Railroad. Pulled by Mikado No. 34, the train is climbing the grade at Red Hill, between South Fork and Jones. Some Sierra trains made the 59-mile run in as little as 4 to 4½ hours, but the average was 5½. Return was made the same day. *(—Photo by Al Rose)*

Sierra Railway No. 20, 2-8-0, westbound on HHRR, passes Oakland Camp in July 1937. Other Sierra locomotives operated on the HHRR at this time included No. 22, 2-8-0, Nos. 30 and 32, 2-6-2 type, and Nos. 34 and 36, 2-8-2 Mikados.

By the end of June 1934, 33,000 ties had been replaced. Work was supervised by engineers from the City and they had several occasions to write letters of complaint. Paul Ost informed the SERA captain in August that the City was dissatisfied with results, as only 80 to 90 of the 600 men were actually employed on railroad work. He claimed that almost two-thirds of the man-days were spent on the camps' own maintenance and concluded that the City was finding costs high in relation to the work accomplished. Nevertheless, the San Francisco *Chronicle* reported that the rehabilitation work had a value of $100,000 to Hetch Hetchy, besides affording sustenance to workers and removing them from many municipal relief rolls.

Hauling these workmen back and forth was an additional expense to the City. Shopmen at Moccasin built six four-wheel flatcars to carry passengers and used them with the small Plymouths to move track gangs around. Railroad men, used to spending most of their time on other work at Moccasin, were all busy while SERA crews were at work. Most of the railroad improvements were finished within six months, two of the camps closing in August 1934 and only three remaining by April 1935. On May 13, 1935, the Sierra Railway started operating on the Hetchy.

The City's own employees were prepared to do the heavy hauling job themselves, but the Sierra won the contract on a bid basis. This was a blessing for steam railroad enthusiasts, because Hetch Hetchy people would have bought two 65-ton Plymouth gasoline locomotives for the job, using them to handle four-car cement trains. Dud Snider prepared a bid for City forces on this basis: seven cents a ton-mile. The Sierra, with a surplus of steamers and everything near at hand, won on a bid of six cents, and it handled the trains for more than three years with efficiency and dispatch – and with a little help from Hetch Hetchy crews in wintertime. Sierra headquarters, dispatcher, roundhouse and shop were at Jamestown, only 15 miles east of the junction. Engines assigned to the Hetchy "branch" laid over at the junction, except when called in for shop service or inspections; dispatching was by phone from Jimtown. Under the agreement, Hetch Hetchy's Power Division retained the privilege of running its own equipment on the rails, but these movements also were controlled by the Sierra Railway dispatcher.

After receiving authority for common carrier operation, the Sierra went to work. Since the short line was then in receivership, this operation probably enabled it to survive the Depression and live on to become the major nostalgia gem it is today. Sierra's large 2-8-2, No. 36, had made a run from the junction to Six Bit Gulch early in 1934 to test a lateral motion device they'd installed. Apparently it was successful, because Mikados Nos. 34 and 36 double-headed the run with six to seven cars at the busiest times. Small-drivered 2-8-0s Nos. 18 and 20, and 2-6-2 No. 32 were also used, often all three together with four or five cars. The Hetch Hetchy people requested engine loads not to exceed 12 tons per wheel, rigid wheelbase not to be greater than 12 or 13 feet, with "blind" drivers required. No cars over 42 feet long or weighing in excess of 60 tons gross would be handled; cars would be fitted with 15-pound retainers to conserve air on long grades. They requested a speed limit of 15 miles per hour to minimize damage to track at curves.

The Sierra operators, for their part, specified they'd accept all freight up to but not over twelve cars per day. They would not be required to handle any train of less than four cars, except in emergency at higher rates. Since they were billing on a weight basis, they named minimum carload weights of 40,000 pounds except cement, 60,000. They agreed to carry passengers at such times of the year as bus service could not operate to Mather, providing a combination car or a motor coach and charging $3.00 one way. Sierra people agreed that when the work was finished, they'd leave the Hetch Hetchy in the same condition as when they started. Since they wouldn't have the old flatcar-mounted push plow, Sierra was to have use of the Hetchy's rotary snowplow – over the strenuous objections of Dud Snider, who claimed Sierra crews had no experience in operating heavy gasoline equipment. The Plymouth was not included in this deal.

Now that the raising of O'Shaughnessy Dam was about to start, the foresight and integrity of Michael M. O'Shaughnessy was seen. He'd built

Work proceeds on the raising of O'Shaughnessy Dam in August 1936. The original structure is shown in foreground with water pouring from the outlet valve at lower right. Concrete is being poured at the other end to bring the new structure out flush with the center, then increase the height by 85 feet. At top of the picture is the aggregate plant with hopper and mixers below. When the work was completed, not a trace of any of these structures was left to mar the landscape. *(—City of S.F.)*

Among equipment brought to Mather by Sierra trains were these steel liners for the 72-inch outlet conduits in the new section of O'Shaughnessy Dam. This was February 1937; snow remains all over the place, but the vital tracks have been cleared by Hetch Hetchy's rotary snowplow.

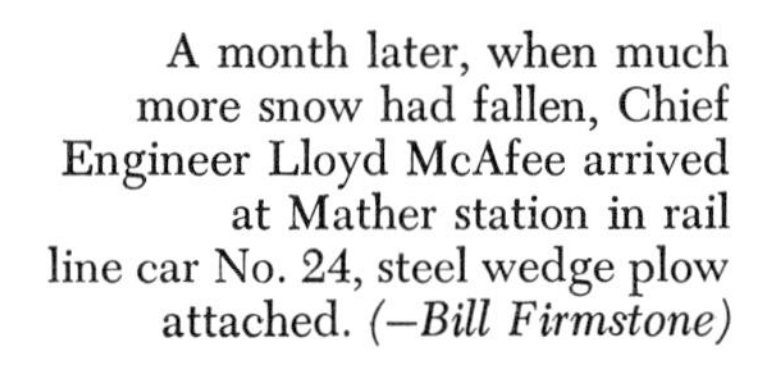

A month later, when much more snow had fallen, Chief Engineer Lloyd McAfee arrived at Mather station in rail line car No. 24, steel wedge plow attached. *(—Bill Firmstone)*

Hetch Hetchy's track truck No. 21 was photographed at Mather in summertime. It had been a favorite passenger railcar during passenger-train days, was converted into a supply truck afterwards. Hetch Hetchy Railroad had the right to operate its own equipment over the railroad while Sierra Railway trains were hauling cement and other supplies for the dam raising, but all dispatching was done from the Sierra's headquarters at Jamestown. *(—Ralph Demoro)*

During the dam raising project conveyor belts were used to carry aggregate materials from the quarry above to the plant at the construction site. Original plans called for a narrow-gauge rail system to do this job, but the conveyor proved to be more economical and was installed. *(—City of S.F.)*

the original dam in such a way that the addition could be made with a minimum of problems and expense. The foundation was sufficiently wide to accommodate the addition; a shelf 80 feet wide had been formed on the downstream toe. A number of other details had been taken care of, at greater original cost, so that the later work would be easier. The enlargement was accomplished by building an addition against the downstream face of the old dam, to increase its thickness, and continuing to a new crest 85½ feet higher.

Such an addition to a structure of the type and magnitude of this dam was unprecedented. It involved special construction features to ensure that immediately upon completion the addition would take up its proper share of the load and the entire mass of the enlarged dam would act under stress in practically the same manner as if it had been originally built to the final dimensions. The original concrete had been cooling, shrinking and aging for 12 to 15 years, so now the new material had to be brought to approximately the condition of the old within the construction period only, to avoid dangerous stresses. This was accomplished by making concrete that was initially of somewhat higher grade than the old when poured and by artificially cooling the concrete as it set.

Construction Engineer for the job was John H. "Buddy" Ryan, who had been on original construction at Second Garrotte and in the Coast Range. Sand for the concrete was dug from a deposit at Miguel Meadow, on the road between Hetch Hetchy and Lake Eleanor, and carried by a three-mile aerial tramway to the aggregates plant at the dam. Rock came from a quarry only a half mile from the mixers, on the hillside behind old Damsite camp. The contractor planned originally to have a rail operation for getting out this rock and delivering it, using four six-ton Plymouths up above and a couple of six-ton Hercules gas locomotives for hauling to the plant. But a six-yard skip on a cableway was used instead, later supplanted by a system of conveyor belts. Cement was carried from a 6000-barrel steel silo at Mather in 15-ton "trailer trucks."

After mixing, concrete was delivered by means of two huge placing buckets pulled through the

Completed and ready for the dedication ceremony in October 1938, enlarged O'Shaughnessy Dam was a perfect union of the old with the new. The tunnel at opposite end of the crest leads to the Lake Eleanor Road, newly elevated above the reservoir line. Right, water is released through outlet valves to flow down the Tuolumne River to Early Intake and the aqueduct tunnel. Below, radial gates of the new spillway at Hetch Hetchy, at the south abutment of the dam, were designed by engineer Bill Helbush; 1967 photo from P.U.C.

Hetch Hetchy Reservoir and O'Shaughnessy Dam as seen from the Lake Eleanor Road in 1961. The old route of Hetch Hetchy Railroad winds up the grade from center to far right; it is now a paved highway bringing fishermen, hikers and sight-seers into the area. *(–S.F. P.U.C.)*

Looking westward in 1961, mountain water pours from the outlets at O'Shaughnessy Dam and falls into the riverbed below for the 11-mile journey to Early Intake. Completion of Canyon Power Tunnel in 1967 took most Hetch Hetchy water out of the streambed, although a sufficient flow is always maintained to keep the canyon beautiful and the fishing good. It might be pointed out that nearly all Public Utilities Commission photos in recent years were taken by photographer Marshall Moxom. *(–S.F. P.U.C.)*

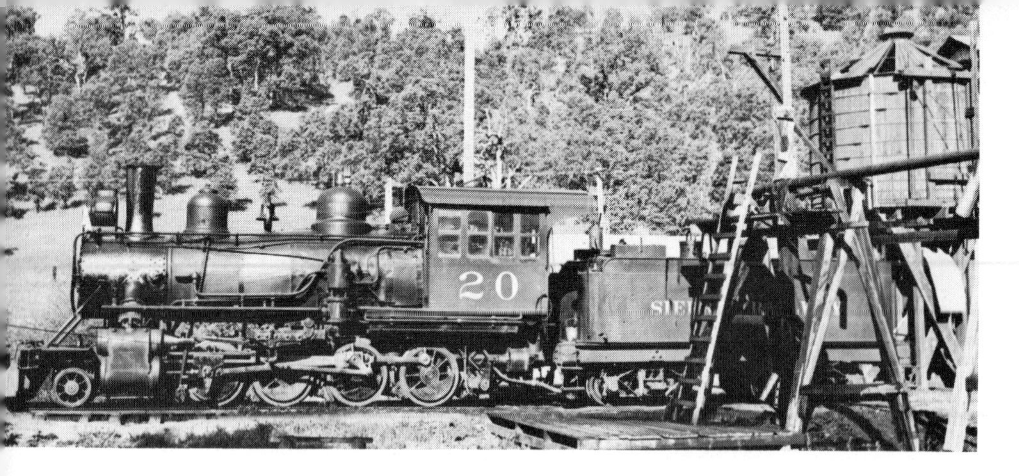

Sierra Railway No. 20, working the Hetch Hetchy Railroad, lays over at the junction on a warm Sunday in May 1937. In back of her was engine No. 32, both of them ready to take a cement train up to Mather the next morning. *(—George Henderson photo)*

air across the 1900-foot canyon on a Lidgerwood cableway with Smilie side control to swing the loads laterally. To avoid interference with the work, no reservoir water could be discharged over the original spillway, so the old by-pass tunnel below was used to keep the water level down. Meanwhile, a new channel spillway, with radial gates designed by Bill Helbush, was under construction in the cliff face to the south. At the north end of the new crest a short vehicular tunnel was excavated so the old Lake Eleanor road could be reached.

Buddy Ryan told of a sentimental thing he had done just before the new concrete was poured. Upon O'Shaughnessy's death in 1934, Ryan had taken the old man's tunnel gear – boots, the large black hat and a special old black coat – from the Livermore warehouse up to Mather. Behind the original dedication plaque of 1923 there was a hollow space. At the appropriate time he had workmen remove the plaque; inside he placed first the boots, toes turned up and soles facing upstream; he folded the coat and put it nearer the opening, with the Chief's hard hat on top. Ryan had the hole sealed up and later watched as concrete was poured over and around it. The plaque was reinstalled above.

So O'Shaughnessy Dam was raised, with all the heavy materials and over 55,000 tons of cement being hauled in over the Hetch Hetchy Railroad by Sierra Railway trains. Operation was on weekdays only and there was usually only one train out on the line at a time. Joe Cavagnero, a locomotive engineer on this operation, said it took more than 12 hours for a round trip and they did it without rest. Joe claimed that either of the Sierra's Mikados could take four loads up Priest Hill with no trouble, even though the line was not in top condition. Jim Baker was another Sierra hogger who worked the Hetchy run, usually on engine No. 32. There was a trip when they put a bunch of cars on the ground just beyond South Fork bridge. The engine and a tank car stayed on, so they cut off and went on to Mather, sleeping in one of the City's cabins. Next morning they met the Sierra's rerailing crew and equipment at South Fork. The whole train eventually got to Mather and the train crew hurried back as far as Groveland for their second night in the mountains.

Snow was heavy in the winter of 1936-37. At the first heavy fall, Sierra crews took rotary No. 1173 and set out eastward from Groveland. They wouldn't take an operator from the Hetchy, said they'd get along fine and engine No. 36 was strong enough for any drift. Dud Snider said it took them a week just to get to South Fork, only 17 miles beyond Groveland and still 15 miles short of Mather. The regulars at Moccasin felt that their rotary would not perform well with a steamer pushing, because the speed had to be kept down to about one mile per hour in fairly heavy snow. Well, the Sierra came back and asked Hetch Hetchy to provide an operator and Plymouth No. 1171, which could handle the slow speeds easily in low gear. They cleared the line and there was no further dispute about who worked the rotary.

At Christmastime, 1936, there was a sudden three-foot snowfall, followed by several days of additional snow. People attempting to return to work at the dam and on Hetchy jobs were stalled in the snow on the narrow, dirt "back road" running from South Fork to Mather. It was reported that 50 to 75 cars were abandoned along the 15-mile stretch, their passengers being brought back to settlements by locomotives and the snowplow on the Hetch Hetchy Railroad.

During 1937 the Sierra Railway of California changed its name to Sierra Railroad as shown here on Mikado No. 34's tender. The engine was photographed at Mather, December 4, 1937, ready to return to Hetch Hetchy Junction shortly after being turned on the wye. Sierra trains stopped operations on HHRR in mid-1938, when dam raising work was completed. *(—Al Rose photo)*

The enlargement of O'Shaughnessy Dam to its ultimate height of 312 feet above the streambed was completed early in 1938. The then-largest dam in California was dedicated on October 16 with speeches by Mayor Rossi, Lewis Byington of the utilities commission and Kenneth Godwin, Regional Director of the federal PWA. Some of Godwin's remarks are of interest today: "This is one of those Public Works Administration projects which must be given credit for being outstanding. It is a notable monument to the wisdom of using public money for improvements which will aid and promote the public welfare rather than to waste our substance on huge preparations to destroy the well being of others as some nations seem prone to do."

By May 1938, when all the contractor's heavy equipment had been moved out and the railroad faced the expense of burning grass and weeds on the right of way, use of steam locomotives was discontinued and the small volume of freight remaining was hauled using Hetch Hetchy's Plymouths, sometimes only east of Groveland. Red Fent said that No. 1171 was on this work for a long time after termination of the Sierra contract, and it was steady — a two-day round trip at least twice a week. Displaying their new name, Sierra Railroad steamers came back for a brief period in July and August, when some heavy train movements were necessary.

Lloyd McAfee sent Julian Harwood to check the City's railroad after Sierra operation came to an end. In June 1938 he reported that the Sierra had been negligent in ditching, in track alignment, in slide repair and in culvert replacement. But a crew of ten men were on the job and would be able to finish the work in three and one-half months, leaving track and roadbed in acceptable condition as specified in the terms of the agreement.

The Sierra Railroad had done well on its bid, turning from net losses in 1932, 1933 and 1934, to earnings of almost $10,000 in 1935, $134,105 in 1936 and $59,066 in 1937. In 1938 earnings dropped all the way to a mere $5.00, but by then the lumber mills were coming back to active life and the Sierra was able to survive into the modern era. At some time in this period Sierra leased the Hetchy's small Plymouth No. 1170 for use in an extensive track relaying job. It could haul one flat and could be left anywhere along the line at night unattended. During the '30s also, the Hetch Hetchy traded a long combination coach for the Sierra's short car No. 6.

When vacated by the contractor, the camp at O'Shaughnessy Dam met full approval of the superintendent of Yosemite National Park. As a result of sand dredging at Miguel Meadow, an artificial lake was created; when the area was cleaned and landscaped, a permanent beauty spot resulted. The quarry south of the dam was also landscaped and became virtually undetectable within the year. It was hard to realize, only a few months later, that more than 400 people had been at work here.

San Francisco's electric power transmission line ends at this switchyard near the Bay west of Newark, just a few miles short of the City. From here Pacific Gas & Electric Company carries some of the power over its lines into San Francisco for such municipal uses as operation of electric buses and streetcars, street lighting and traffic signals. Other power is disposed of to irrigation districts at Turlock and Modesto and to a few industrial plants in the Bay Area. *(–S.F. P.U.C.)*

23

END OF THE LINE HERALDS A NEW ERA

Having completed the four-year job of preparing for and raising O'Shaughnessy Dam, the Power Operating Division of Hetch Hetchy settled back into its quiet country living in scenic Tuolumne County. At Moccasin headquarters the powerhouse whirred noisily around the clock; personnel were busy in offices and shops; homes were added as the number of married employees increased. And the powerhouse operator on duty in the big, glassed control room continued dispatching trains for the railroad as needed. For it was still a time when mountain roads were not plowed out after winter snowstorms. In fact, making a phone call from Tuolumne County to San Francisco could still be a chore.

The village of Moccasin has been a little piece of the City stuck out in the Sierra foothills 18 miles from Sonora. About 100 people live there today, City employees and their families, but in the early years many of the workers were single men. About 30 of them lived at the clubhouse and in a near-by dormitory, taking their meals at the big clubhouse dining hall where cook Mickey Ryan carried on the tradition of good food. Mary Conty was housekeeper; she delighted in having important visitors from the outer world to accommodate, especially if she could prepare a special treat of her native Italy. That was Moccasin: the powerhouse, the clubhouse with its offices, young Bill Segale's store, the railroad, the country atmosphere, the notorious summer heat. There were just enough native Poison Oakers and Sonorans working there to keep it from getting too citified.

It's not the same today. The isolation of the past is gone and over new roads Moccasin is almost a suburb of busy Sonora. Those years of relative isolation, though, were the times of happiest memories for families stationed there — the people that made up what amounted to one large close-knit family of Moccasin. They had their own school, and the children were taken on their annual picnic to Mather over the railroad in track bus No. 24. When a gold dredge was working the Tuolumne River streambed between Stevens Bar and Jacksonville later in the '30s, big oak trees were shoved over and left for Moccasin residents. Bud Meyer said they would get together and ride four or five miles down the railroad on the big Plymouth and a few small flatcars. All would pitch in sawing up the trees on evenings and week ends, then hauling the firewood back to their little settlement on the "special train." Some of the high school boys had permission to head east on a Buda speeder and shoot their bucks on opening day of deer season.

There were always a few men working on the railroad, again keeping it in minimum condition for winter and with an eye on planned construction work. Early Intake and Moccasin inclines were kept in good condition and used often, the latter on maintenance of the penstock, especially after the hotter days when leaks sometimes developed in various expansion joints. Auto machinists would be sent up the incline to do this repair after midnight so the line could be ready for the "morning rush." Ties were replaced, trestles strengthened, rollers repaired — the tramway lasted through 1957. Shopmen were busy in 1939 completely rebuilding the body of rotary snowplow No. 1173.

Famous among the working crew was Eddie Webb, early-day stagecoach driver on runs out of Coulterville and Chinese to Yosemite. According

In 1940 the late Bill Pennington photographed the old station at Hetch Hetchy Junction with the ascending grades of Sierra Railroad to the left and Hetch Hetchy to the right. Some station buildings had been demolished a year or so before, but this last relic was destroyed in a fire about 1947. *(—Bill Pennington photo)*

An interesting character on the Hetch Hetchy payroll at Moccasin was former stagecoach driver Eddie Webb. His hobby was searching out and restoring old coaches for use in movies and local parades. This 1936 snapshot shows Eddie with one of his coaches and teams in near-by Columbia. *(—Cliff and Mona Hanson)*

By mid-1948 there was little activity on Hetch Hetchy Railroad. The line was occasionally cleared of fallen rock and trees so that snowplow service could be provided in wintertime. During summer the unused track sat and baked in the hot sun; travelers on the highway just assumed that it had been abandoned. Above, the yard and warehouse at Moccasin (powerhouse in background). Right, trestle over Middle Fork near Oakland Camp.

Homemade rotary snowplow at Groveland in 1947. Formerly using the body of caboose No. 16, the car had been renumbered 1173 and a new body constructed in the late '30s. It was thirty feet long, divided into a motor compartment and living quarters; the radiator on top cooled water for the huge engine inside. *(—Bert Ward photo)*

All activity had ceased at Hetch Hetchy Junction when this picture was taken in 1940. In center are foundation blocks and turntable for the once-busy gasoline railcar roundhouse. Looking eastward, Sierra Railroad is in background. *(—Bill Pennington photo)*

On May 25, 1947, Hetch Hetchy Railroad held its first and only excursion for railroad enthusiasts, but only on the portion from Groveland to Mather and return. Roy Graves remembered that it was a very thrilling ride with unrivaled scenery all the way and plenty of photo stops like that below on the cliff just west of Mather. *(—Roy D. Graves photo)*

to numerous old-timers, he was the most beloved of all Moccasinites. Hank Femons recalled that Webb, of Indian and Negro descent, was known as "low-gear Eddie" when he drove the garbage truck around the village — he never bothered to shift gears! But then, Eddie Webb wasn't interested in motor vehicles; he collected and operated stagecoaches. On evenings and days off this practical enthusiast would search for old coaches, haul them back to Moccasin and restore them to operating condition. Red Fent says he often helped with this work in his spare time, cutting leather to make strap springs, for instance. Tom Hope helped with blacksmithing. Two or three of Webb's stages were used by film companies every summer; he rented out coaches and teams and became a great friend of numerous film personalities. Eddie Webb was one of the highlights of early Moccasin.

The sole reason for Moccasin's existence, of course, was the big powerhouse, turning out 80,000 kilowatts (100,000 hp) of electric power every day. By keeping this plant in full production San Francisco was getting back some of the money it had spent on Hetch Hetchy. When the 98½-mile transmission line was extended from Moccasin to San Francisco Bay in 1925, no definite plan for disposal of the energy had been found. After much difficulty and some unpleasantness, a day-to-day contract was entered into with Pacific Gas & Electric, whose Newark Substation happened to be adjacent to the western end of the transmission line. This company bought the entire output of Moccasin, less such energy as was required by the City. Voters in 1937 rejected a plan to extend the City line into San Francisco, while also turning down over the years several proposals to take over the power system as they had done so successfully with the water.

It was on this power line that the late Art Crowley, in charge of its maintenance for many years, with his crew developed methods and tools which were to make San Francisco's transmission line famous — tools that are now used all over the world. Previously, it had been necessary to cut off current while repairs were being made or improvements installed. No one in his right mind would go anywhere near a 115,000-volt line while it was "hot." But Crowley's tools and methods made this an everyday job, saving millions of dollars through avoidance of shutdowns and virtually eliminating accidents. The equipment was designed and built up at Livermore, when that was maintenance headquarters for the transmission line before transfer to Moccasin in 1954. Working like a well-trained football team, Crowley's crews could do in one-half hour what had formerly taken a whole day's work.

Over the years there was a steady, bitter, hard-fought battle going on between the City and County of San Francisco and the United States Department of the Interior over just how the City would comply with conditions of the Raker Act, under which it was specifically forbidden to sell any produced power to a private utility. Secretaries of the Interior Ray Lyman Wilbur under Hoover and Harold Ickes under Roosevelt fought San Francisco. In fact, all through the years from 1923 to the start of World War II the argument went on, with the City offering a series of different proposals for power distribution and the Interior Department saying No! The Attorney General in Washington actually instituted suit against the City in April 1937 and won, and the City appealed.

In April 1940 the United States Supreme Court found San Francisco in violation of the Raker Act, bringing about a vote in 1941 on acquisition of P.G.&E.'s City distribution system. This was dubbed "Plan Nine"; in presenting it Utilities Manager Cahill predicted that if it failed, the federal government would probably seize Hetch Hetchy power and operate the plant in conjunction with the Central Valley Project. Thoroughly brainwashed by private utility propaganda, the citizens again voted against public ownership, so speedy efforts were then concentrated upon amendment of the Raker Act. At this point, President Roosevelt had something to say, writing that the act had been passed and the project developed upon definite assurances that the benefits if conferred "would not be alienated from the people to private interests. . . . For more than a generation this condition to the grant, agreed to by San Francisco, when it accepted the privilege of developing a water supply and source of hydroelectric power in a national park, has been violated and the people of San Francisco have not received

To maintain Hetch Hetchy's power transmission line running from Moccasin to Newark, special tools were designed and constructed in workrooms and shops at Livermore and Moccasin so that the line work could go on without shutting off power – something that had never been done before anywhere. Art Crowley was in charge of this work; the special tools were designed under his supervision. He is shown, right, holding his "armor rod" in 1946. Above, the crew works a live line on one of the transmission towers near Moccasin. Upper right is an earlier picture showing Art Crowley installing vibration dampers on a dead line in 1932, with the orchards of San Joaquin Valley far, far below. *(–Three photos, Art and Maude Crowley)*

These three photos were taken in 1940 by Doug Mirk, then Chief Operator at Moscasin Powerhouse. They show maintenance work on one of the four generator units. After the top section was removed (center), the generator with attached Pelton water wheels was lifted by the building's overhead crane, as shown at top. In operation, high-pressure water from the penstock is directed at the water wheel buckets, turning the unit at high speed to generate electricity. Below, a welder makes repairs to one of the buckets. *(–Muriel Mirk)*

Hetch Hetchy road trucks Nos. 106 and 108, built by White Motor Co. in 1920, and Nos. 120 and 121 of 1924, were still in service as late as 1948. Here, with a little help to balance the weight, one of them moves a heavy cable spool in 1939. *(—Art Crowley photo)*

The winter maintenance equipment for snow country east of Groveland consisted of former railcar No. 24 and Plymouth loco No. 1170, shown here at Groveland in 1947. No. 24 had its motor removed and a four-wheel truck substituted at the rear. The 1170 shoved it through the lighter snows with line crew and occasional supplies inside. *(—Dave Welch photo from Marvin Maynard)*

On a couple of Sundays in 1947, railfans with their own track velocipedes and speeders inspected the upper portions of Hetch Hetchy Railroad with official sanction. Warren E. Miller to left; Stanley Snook to right. *(—Bert Ward photo)*

The railfan inspection party of 1947 pauses at Intake station, at the top of Early Intake cable incline. At left is Yosemite Valley Railroad speeder M-16, owned by Al Rose who is shown standing on the station porch. *(—Bert Ward photo)*

the benefits that Congress expected them to get." He could see no prudent Congress or administration amending Raker and was confident that a feasible solution would be worked out within the law, "thus conserving the resources of the national parks for the people."

Pearl Harbor saved the City this time. A defense aluminum plant hastily started up at Riverbank legally took the disputed Hetch Hetchy power output. However, another crisis arose when the plant closed late in 1944, and the United States gave the City six months to comply with the act and with court decisions. Thus, in 1945 San Francisco negotiated new contracts for disposal of Hetch Hetchy power. These contracts, with Modesto and Turlock Irrigation Districts and the private utility, were approved. In 1960 and 1966 additional agreements were negotiated to provide a legal market for power generated at the new Holm and Kirkwood powerhouse, about which more later.

The City's three power plants now generate approximately two billion kilowatt-hours of electric energy a year, gross revenues averaging about $13,000,000 annually. There are groups currently active in San Francisco which offer documentation to claims that San Francisco still fails to comply with terms of the Raker Act, that the private utility profits greatly through arrangements to deliver Hetch Hetchy power into the City municipal plant, that the people pay higher rates than they would have to if they were able to use their own electricity. This takes into consideration that the utility pays taxes and so on. If the Water Department is any criterion — and the same arguments were used against purchase of that private utility — the natives are being cheated and all the great potential of Hetch Hetchy is not being realized. However, that's another story.

More development work was still to be done in the mountains. Seventeen miles northwest of Hetch Hetchy lay Cherry River and an area planned for reservoir use from the very first. Starting in 1939, the City spent $25,000 each year for eight years on surveys, studies and plans for development of this water resource. This provided an additional reason for maintaining the Hetch Hetchy Railroad through and after World War II until such time as it "will again be required for heavy freight haul." The line was kept open as "an industrial spur" railroad.

But the old rail system was gradually disappearing. The station and other structures at Hetch Hetchy Junction were removed in 1937-38; the old sand house and oil tank at Groveland were dismantled in 1944; many spurs and sidings were removed or cut back to reclaim tons of steel during the war years. Still, there was always a little construction work in progress: 1850 ties in 1944; two bridges repaired, a new gas pump at Groveland, fences and cattle guards repaired in 1945; a new spur in Moccasin yard in 1946. Snowplowing crews were out as much as a week at a time, though there were more frequent derailments of the rotary as track deteriorated in the late 1940s. Then, as back roads in the mountains got paved and better highway snowplows were developed, it became obvious that the railroad was an ex-

Mather station in May 1947, with the Railway & Locomotive Historical Society excursion train ready for the return trip to Groveland. A year later the station was gone and track removal started. *(—Dave Welch photo from Marvin Maynard)*

pensive luxury. As for future hauling for a Cherry River dam project, this would require a long truck haul from the railroad in any event and new road equipment was capable of doing the entire job. Incidentally, some of the Hetchy's 1914 and 1916 White trucks were still in service through the Second World War.

Railfans discovered Hetch Hetchy just before the end. In May 1947 a group was able, after much negotiation, to charter a final passenger train: locomotive No. 1171 and coach No. 6. They ran from Groveland to Mather and return, 64 miles, most of them enjoying the marvelous scenic beauty of this railroad for the first (and last) time. Others had private "fan trips." Al Rose and Stanley Snook received permission to operate their private speeders and track velocipedes on certain Sundays. Historian Bert Ward was along on these runs; he was impressed with the steep grades, sharp curves and rugged scenery along the old route. Just in time they were, for it had been decided that the railroad right of way above South Fork would make a fine base for a needed road, both to serve for maintenance as far as Mather and to transport material to the Cherry project.

People living at Moccasin were upset when railroad abandonment was proposed. They fought it because they had no other handy way to keep in touch with the outer world at times when the highway alongside the Tuolumne River underwent periodic floodings near Jacksonville. In 1949 San Francisco called for bids for sale of track material and equipment of the railroad. The high bid of $45,680 was termed "grossly inadequate" by the Public Utilities Commission. It was rejected and a negotiated sale held later in the year netted $147,500 from the Purdy Company. One scrapper said the rail was in excellent condition, showing very little use; 7000 tons were sent to be used in South America and India, along with a few pieces of equipment. Most of the 21 freight cars, the two cabooses and combination coach No. 11 were worthless and contributed to a huge bonfire at Moccasin when the line was dismantled. Railcars Nos. 21, 23 and 24 were scrapped; No. 19 was bought by rail enthusiast and historian Al Rose for preservation. Three Plymouth locomotives and three Buda speeders were sold; A. D. Schader Co. bought the small Plymouths, and twenty-ton No. 1171 eventually wound up as a switcher on the Parr Terminal Railroad of Richmond, California.

By the beginning of 1950 the Hetch Hetchy Railroad was just a memory, although two short pieces of track remain as monuments: those embedded in concrete over a shop pit at Hetch Hetchy Junction and those in the concrete floor of old Moccasin Powerhouse. The last items sold for scrap went in February 1951: a Groveland shop building, wooden trestle and water tank. Proceeds: $652. It is interesting today to note in passing that the Hetch Hetchy Project became so vast and had so many outstanding engineering features that the building and 34-year operation of a 68-mile, standard-gauge mountain railroad is today not even celebrated or mentioned as one of the accomplishments.

Combination coach No. 11, obtained from the defunct Ocean Shore Railroad at San Francisco, saw very little service after 1925. Here she sits on the side track at Moccasin, March 1946, waiting for the end. Passenger-train cars were painted olive green on the Hetch Hetchy.

Track Truck No. 21, her years of service ended, was in storage at Moccasin in August 1949. Track removal was already going on and No. 21 would in a few months be junked for scrap.

Last passenger train on the Hetch Hetchy was this May 1947 excursion of the Railway & Locomotive Historical Society. No. 1171, the "big" Plymouth, is shown with coach No. 6 on the return trip to Groveland from Mather. On account of the long runs required of this engine, extra fuel and water tanks were adapted from 55-gallon drums and mounted on both sides of the hood. This light equipment was ideal in the final years, considering the deteriorated condition of the roadbed. *(—Dave Welch photo from Marvin Maynard)*

Old No. 19, the original track bus-railcar-ambulance of 1919, sits in retirement thirty years later at Moccasin. Tracks were already being removed by the scrappers. Fortunately, No. 19 was saved through the efforts of railfan, author and photographer Al Rose, who purchased the car and had her stored at a museum near Modesto until such time as Yosemite National Park found a place for her in the transportation museum at El Portal. And there she sits today, with Hetch Hetchy Shay No. 6. *(–Drawing by Irene Brown)*

Parts of the old right of way were used for the new City-built road to Mather, and in making State Highway 120 (Big Oak Flat Road) an all-year access road to Yosemite. Little remains in situ for the railfan and industrial archacologist, but the old route can be seen and hiked all the way up Priest Hill, and much is visible along the highway beyond Groveland. At the latter place the body of rotary No. 1173 served as someone's garage for years. There are trestle remains at Big Creek and South Fork; the Intake incline skip was left grounded beside the old winding house. Four of the six steam locomotives owned by Hetch Hetchy have been preserved. Two are at Travel Town, Los Angeles, bearing names of subsequent owners; one is at Yosemite; the fourth is in Oregon. On display at Yosemite with engine No. 6 is the old Chief's favorite track bus, No. 19. Its mechanical parts were restored to good operating condition by the White Motor Company.

Hetch Hetchy has continued to grow since its old railroad artery-backbone disappeared more than twenty years ago. Big projects have been completed up in the mountains; major co-ordination and co-operation has been worked out with Modesto and Turlock Irrigation District people; the Tuolumne River has been completely "tamed." One minor but notable work was the installation of wooden shutters on the cookhouse at O'Shaughnessy Dam "as protection against the entry of bears." The most important job of all was getting the City and the irrigationists together. According to hydraulic engineer Max Bartell, the irrigation people referred to them as "Those Hatch Hatchy [*sic*] sons of bitches!" Bartell was the ideal man to work at getting these two factions together: he'd made population studies upon which Hetch Hetchy operations were based; he'd been a student of human nature since his dad had sent him as a teen-ager to live with Indians in the '90s.

At the direction of Mike O'Shaughnessy, whom he intensely admired, Bartell had been sent early in the 1930s to see what could be worked out with the suspicious citizens at Modesto and Turlock. The idea was to change from fighting to finding areas where they could get along and perhaps work together. Following a visit by O'Shaughnessy and City Attorney O'Toole to Turlock, a simple agreement was signed in 1934 by Roy Meikle, Chief Engineer at Turlock, and Max Bartell for

Death came to Hetch Hetchy Railroad when the last rail was removed a day or so after the end of 1949. The upper left photo shows rail joints being unbolted near Moccasin, using a Hetch Hetchy speeder with generator and electric impact wrench. In center is some of the scrapper's unusual equipment as the last mile is approached near Hetch Hetchy Junction. An old Army half-track draws the pushcar of rails. In background can be seen one of the small Plymouth locomotives fitted for light crane service. Fifteen years later all that remained were rotted ties and tottering bridge timbers, the former on the grade near the junction (upper right) and the latter at the crossing of Big Creek (lower left). *(—Scrapping photographs by Stanley Snook)*

Disposition of Hetch Hetchy Railroad's steam locomotives is of interest mainly for the unusual fact that four out of the six of them remain in existence! Only No. 1 and No. 3 have been scrapped. The former 1-spot is shown, left, in 1937 in the shop of Pickering Lumber Co. near Sonora, where she operated for many years as their No. 7.

Engine No. 3 was sold in 1927 to the California & Oregon Coast Railroad, where she operated out of Grants Pass as their No. 301. *(—1938 photo by Doug Richter)*

Heisler No. 2 was sold along with No. 1 to Pickering in 1923. She retained the same number on the logging railroad east of Sonora. She is shown here in 1949 at the Pickering mill, Standard, Calif. The engine has been preserved at Traveltown, Los Angeles. *(—Doug Richter photo)*

The Hetchy's big Mikado, No. 4, was sold in 1927 to the Newaukum Valley Railway in Washington State. She later served on the Santa Maria Valley Railroad. *(—Warren E. Miller, Railway Negative Exchange)*

Hetch Hetchy's old No. 5 is preserved on display at Sutherlin, Oregon, after putting in many years' service as No. 100 on railroads of the Weyerhaeuser Timber Co. She is shown at the mill, Springfield, Ore., in 1961.

Former No. 4's last operation was on the Santa Maria Valley Railroad in California. A very attractive locomotive, this machine is on permanent display at Traveltown, Los Angeles, not far from her old sister, No. 2. *(—Doug Richter photo)*

Last steam engine acquired by Hetch Hetchy, Shay No. 6, went to Pickering Lumber in 1926. As the lumber company started to cut back its railroad logging operations, No. 6 was donated to Yosemite National Park in 1960, for display at the museum at El Portal. Between Mather and Damsite, Hetch Hetchy Railroad was actually within the park borders. The photo was taken in December 1960 near Mariposa, as the 6-spot was en route to the museum. *(—National Park Service photo from Hank Johnston)*

Planning for development of a reservoir on the Cherry River started in 1940, with the City and irrigation districts of Turlock and Modesto in co-operation. Max Bartell, left, from San Francisco and Roy V. Meikle, Chief Engineer at Turlock, are at Cherry damsite, October 15, 1940.

the City. Jim Turner, head of Hetch Hetchy from 1942 and later Manager of Utilities, said that Bartell conceived the idea of getting city fathers and irrigation officials together to talk about co-operation on the Tuolumne. Valley people were brought up to the picnic grounds at Stone Dam on the Peninsula; after a few drinks, all was friendliness.

There followed a series of agreements starting in 1939, each a little more bold and specific, eventually ending in complete accord and hearty co-operation for the entire development of Tuolumne waters. It was in 1949 that four interested agencies signed agreements: San Francisco, Modesto Irrigation District, Turlock Irrigation District and the federal government (through its interest in flood control). The City was to immediately construct a reservoir in Cherry Valley, and the districts would at a later date develop a new, larger reservoir at Don Pedro; they would assist in the financing at Cherry and San Francisco would later aid them at Don Pedro. For flood control benefits the United States would pay about $14,500,000 toward both projects. The City thus arranged for future water storage at about one-fourth the cost of building several City-owned reservoirs at high mountain locations.

After timber had been cleared and sent out on narrow-gauge logging trains of the Westside Lumber Company, work on the Cherry Valley Project commenced in 1950 under the direction of Oral L. Moore, Construction Engineer and later General Manager of Hetch Hetchy. Completed in 1956, the huge Cherry Dam is a composite earth and rock embankment 2600 feet long, with a height above bedrock of 330 feet and a base thickness of 1320 feet. A six-mile power tunnel and penstock carries water to Dion Holm Powerhouse on the Cherry near its confluence with the Tuolumne River. A mile-long tunnel was later driven to connect Cherry Lake with Lake Eleanor, enabling runoff from the latter's watershed to supplement storage at Cherry.

Another power-producing facility was shortly added near O'Shaughnessy Dam. For years Hetch Hetchy water had been allowed to run 11 miles down the streambed to Early Intake; now a tunnel would carry it from the reservoir. Planned from the start for later completion, this Canyon Power Tunnel was put inside the north wall of the river canyon rather than the south. Finished in 1965, it delivers 550 m.g.d. to the Robert Kirkwood Powerhouse, in the canyon just upstream from Early Intake Diversion Dam. This plant and near-by Holm Powerhouse on the Cherry are normally operated by remote control from Moccasin, some 20 miles to the west (to the mountain traveler it seems much, much farther). Power from these plants leaves Intake Switchyard on a new steel-tower transmission line of 230,000 volts, is carried 48 miles to Warnerville Substation near Oakdale and eventually reaches the irrigation districts.

In 1969 a new powerhouse of the modern outdoor type replaced the old building at Moccasin. A year later the exposed steel pipe joining the two sections of the Mountain Division Tunnel down in the canyon below South Fork was replaced by an underground siphonlike section, thereby eliminating a weak link in the aqueduct chain.

Then the new Don Pedro job got under way with a massive rock-fill dam about a mile and a half downstream from the old Don Pedro Dam of 1923. Builders of the latter would never have imagined in their wildest dreams that less than

Last piece of railroad equipment in the Hetch Hetchy area was the body of rotary snowplow No. 1173. Someone at Groveland used it as a garage for many years. Shown here in 1965, it was still standing at the time of publication of this book in 1973.

With the abandonment of the railroad, Hetch Hetchy's Sno-Cats took over the job of getting linemen and maintenance men out to trouble spots in winter snows. Here, just west of Mather in 1950, are Pete Golub, Art Crowley, Les Phelan and Carl Seward, ready to tackle repairs on the transmission line. *(—Art Crowley)*

Below, the earth- and rock-fill dam at Cherry Valley nears completion in 1955. *(—S.F. P.U.C.)*

Construction of Cherry Penstock, leading to Holm Powerhouse, looking southwest from Granite Portal in 1959. The view to left shows work being completed in Canyon Power Tunnel above Kirkwood Powerhouse in 1964. *(—Both, Marshall Moxom photos from S.F. P.U.C.)*

Work on Canyon Power Tunnel in 1964. This tunnel was designed to carry water from Hetch Hetchy Reservoir to Kirkwood Powerhouse at Early Intake, thus by-passing the old flow of Hetch Hetchy water down the bed of the Tuolumne River. *(—S.F. P.U.C.)*

Aerial view of Cherry Dam and Cherry Lake, a few miles north of Hetch Hetchy. Water from this reservoir generates power at Holm Powerhouse, then flows many miles down the Tuolumne River into Don Pedro Reservoir. *(—S.F. P.U.C.)*

Construction of the new powerhouse at Moccasin in 1968. Of modern design, it is able to generate more electricity using the same amount of water flow and the same old penstock pipes shown in background. A corner of the old powerhouse appears at left; it has been left standing as part of an attractive view across the lake from Highway 49.

With completion of a new Don Pedro Dam, water began filling a tremendously larger reservoir area in 1971. View above, from relocated Highway 49/120 toward former site of the old Gold Rush town of Jacksonville in August 1972, shows old railroad grade to left of stream. Left below, looking up Tuolumne River toward Stevens Bar, the new highway bridge leading to Jamestown and Sonora stands high and dry. A year later the railroad grade was covered and water stood halfway up the bridge piers.

Two new powerhouses have been added to Hetch Hetchy in recent years. Dion Holm Powerhouse (right) was placed in service on Cherry River in 1960 and Robert Kirkwood Powerhouse (above), on Tuolumne River near Early Intake in 1967. *(—Right, S.F. P.U.C.)*

The lake at Hetch Hetchy, a lovely sight for most of the year, lies clear and smog-free in July 1967, while crowded Yosemite Valley, a few miles to the south, is so cluttered with haze that pictures of the lovely sights are almost impossible. Behind Hetch Hetchy are miles of fine, uncrowded trails, provided by the City of San Francisco under terms of the Raker Act. *(—S.F. P.U.C.)*

half a century later it would lie 200 feet below the surface of a new reservoir with a dam 580 feet high and 800 feet across. Old-timers — Max Bartell for one — would have argued that the capacity of two million acre-feet would never be attained. But the job has been completed and the reservoir line extends back 24 miles into the foothills — far beyond Jacksonville, which exists no more, up the canyon beyond Ward's Ferry Bridge and up Moccasin Creek almost to the powerhouse! The old Hetch Hetchy Railroad route, laid down here in 1916, will be below high-water level from Red Mountain Bar to the Priest Hill grade.

With new Don Pedro the increase in storage capacity will assure the adequacy of San Francisco's water supply to meet its estimated requirements beyond the end of this century. If Don Pedro had not been pushed, fought for and built, it is conceivable that San Francisco would have had to make cutbacks in supplying water to its

When Cherry Dam was constructed and other additions made to the Hetch Hetchy system in the 1950s and '60s, there was no railroad to do the heavy hauling. Trucks took over the work of bringing in cement, aggregates, parts and supplies — and they did a good job, although residents of Groveland noted that they were much noisier than old Hetch Hetchy Railroad. Here, an empty aggregate rig climbs the grade out of Early Intake in 1965.

suburban neighbors to the south. The City now speaks of enlarging Lake Eleanor to the capacity originally planned and approved, but before even a statement of definite plans has been made, environmental opponents are preparing for battle. There exist several other plans to increase the capacity of Hetch Hetchy when and as needed. Many will have little or no noticeable effect on the environment; others will. Tuolumne County residents, Sierra Clubbers and other outdoor types seem prepared to fight each as it comes up for consideration.

Bill Reed, retired Chief Real Estate Appraiser for San Francisco, said that he considered Hetch Hetchy "one of the greatest things that was ever handed to the people of any city" and he hoped that some day they would come to realize this. Jim Turner pointed out another largely unappreciated benefit accruing from the way Hetch Hetchy was designed and built. Even though water rates are still below those in effect in 1930, when the private company was taken over, Hetch Hetchy has no taxing power and gets no tax support whatever. Other systems look great financially, but they whack the taxpayers to achieve this; consumers subsidize their water districts through taxes and still pay higher rates than San Franciscans. Hetch Hetchy water is also furnished to City and County departments without charge; electricity for municipal purposes is supplied at cost. For the taxpayers, these factors amount to a saving of over two million dollars per year.

It looks as though all the arguments advanced sixty years ago and more against damming Hetch Hetchy are purely academic in the light of today's reality. All water of the Tuolumne is conserved; so is that of the Mokelumne, which had been held up as an alternative. "Ruined" Hetch Hetchy Valley is today one of the clean, smog-free places in the national park, quiet, peaceful, a joy to visit. One faction of the Sierra Club wrote in the 1960s of the "needless dam and an irretrievable loss" at Hetch Hetchy. Hardly needless; possibly more gain than loss; and certainly not irretrievable. When mountain storage of water is no longer needed, restoring Hetch Hetchy to much of its original appearance would take perhaps ten years. Meanwhile, and until then, it has been preserved for future generations to enjoy.

Roy Meikle, long-time Chief Engineer of Turlock Irrigation District, said in 1965, "After more than 50 years, I today wonder how a group of lawmakers could have ever got out such a good law as the Raker Act." A great many present-day planners and engineers agree with this opinion. They say San Francisco went to the right source for water and developed it in a manner best suited to the times with the machinery then available.

As motorists use all-year Highway 120 today, to and from Yosemite, they little realize that the "ancient" Hetch Hetchy Railroad provided part of the roadway, the Hetch Hetchy Project provided tunnel granite for the modern road surface, and the scenic beauty along the route is largely the result of San Francisco's care in using and living with nature in the mountains. The ghosts of a score of construction camps haunt the way. They speak of happy days forty, fifty and sixty years ago when the "Spirit of Hetch Hetchy" was alive in thousands who lived and worked on the project. Hopefully, something of that dedication has been captured in these pages.

APPENDIX 1

LIVES LOST IN HETCH HETCHY CONSTRUCTION
(Incomplete)

Peter G. Achatz	Caught between cement bucket and form at dam, 10/12/36
Mike Aracic	Knocked off dam at Hetch Hetchy, 2/24/23
James Bailey	Fell from cable at Hetch Hetchy, 4/17/23
Leonard S. Bailey	Struck head while riding cables at dam, 10/24/35
William John Baker	Skull fractured in accident at dam, 11/22/35
K. Bondarinka	Crushed at Moccasin by locomotive No. 4's pilot, 4/5/24
Erick Boyd	Crushed under avalanche, 2/9/22
Martin Brett	Killed by dynamite explosion, South Fork Tunnel, 3/24/20
George A. Carlson	Fell from dam at Hetch Hetchy, 1/28/23
Pat Carroll	Collision on Hetch Hetchy Railroad, 2/11/22
Ed. Castraine	Caught in revolving belt and pulley, 11/26/21
C. R. Cavanaugh	Mitchell explosion, Coast Range Tunnel, 7/17/30
Claudio Cendon	Crushed by boulder, roadwork at Hog Ranch, 1/9/15
Mike Cheski	Fell to foundation of dam at Hetch Hetchy, 12/1/21
J. Coler (a.k.a. H. E. Kaub)	Mitchell explosion, Coast Range Tunnel, 7/17/30
S. Cotter	Struck by rolling rock at Damsite, 9/12/21
James N. Covington	Skull fractured at dam, 10/24/36
Lee R. Crowley	Mitchell explosion, Coast Range Tunnel, 7/17/30
William Davis	Run over by tram car at Moccasin, 2/19/25
Amico De Iullo	Fell from dam at Hetch Hetchy, 4/7/23
Ed. Donming	Tunnel cave-in, 4/28/21
Joe Dower	Fell from dam at Hetch Hetchy, 2/22/23
Arthur Everart	Fell from steel tower, 12/23/29
Michael A. Fitzgerald	Dynamite explosion at Hetch Hetchy, n.d.
Thomas P. Fleming	Runaway train at Six Bit Gulch, 9/12/22
Roy Ford	Falling rock at Early Intake, 7/26/24
Thomas Ford	Explosion, Priest Tunnel, 1/9/22
F. Foster	Suffocated, gravel bin at Moccasin, 7/14/29
Pat Gallagher	Mitchell explosion, Coast Range Tunnel, 7/17/30
Felice Gastaldo	Shifting pipe on train at Cavagnero, 1/19/24
Charles Goodwin	Falling blocks at rock crusher, Early Intake, 1/17/25
Mike Haley	Explosion, Priest Tunnel, 6/15/22
Telesphore Jannard	(Chute tender) Fell down chute, 8/22/26
E. W. Jennings	Mitchell explosion, Coast Range Tunnel, 7/17/30
George A. Jones	Accident, autostage 96, on divide near Groveland, n.d.
A. Juiz	Struck by derrick boom, 7/7/22
Arley Kellam	Drowned in accident at dam site, 5/3/21
Pat Kelly (Kelley)	Thrown from speeder on Hetch Hetchy Railroad, 3/4/22
P. H. Kilker	Accident at dam, Hetch Hetchy, 8/4/21
Con McBride	Foothill Tunnel accident, Hetch Hetchy Junction, 2/12/28
Pat McCarthy	Struck on head by concrete bucket at dam, 5/8/21
Joseph McGowan	Fell from Hetch Hetchy Railroad train, 8/29/17
Patrick H. McGuire	Fell from freight car at Groveland, 10/11/21
J. M. McMaster	Mitchell explosion, Coast Range Tunnel, 7/17/30
John McNichols	Mitchell explosion, Coast Range Tunnel, 7/17/30
Tony Marich	Skull crushed working at Hog Ranch, 1/7/15
John J. Marshall	Rock fell on mess hall, South Fork Camp, 11/11/22
J. E. Maybin	Mitchell explosion, Coast Range Tunnel, 7/17/30
Frank Miller	Explosion, Priest Tunnel, 1/9/22
Ethel Earl Moyer	(Nurse) Hospital fire at Groveland, 7/27/22

G. Nations	Mitchell explosion, Coast Range Tunnel, 7/17/30
H. K. Nickals	Fell from dam at Hetch Hetchy, 2/10/23
Henry Niemi	Tram accident between Priest Portal and Moccasin, 2/20/25
Joe Norak	Dynamite explosion, South Fork Tunnel, 3/24/20
George A. Parker	Struck by rolling log, 4/13/23
Leonardo Ramiero	Drowned at Hetch Hetchy, 7/16/22
Tony Redka	Mitchell explosion, Coast Range Tunnel, 7/17/30
D. O. Robinson	Struck by cable, 5/3/20
John A. Sandahl	Struck by lightning at Hetch Hetchy, 5/23/21
Angelo Thomas Segale	Explosion, Priest Tunnel, 6/15/22
Joe Sheldon	Struck by tunnel train at Mitchell Shaft, 12/18/31
Dick Sladden	Explosion, Priest Tunnel, 1/9/22
Harry Mosley Snail	Fell from truck at Moccasin, 9/20/26
Horace C. Thompson	Fell from concrete tower, Hetch Hetchy, 9/11/22
C. Urich	Mitchell explosion, Coast Range Tunnel, 7/17/30
Edward Whelan	Fatally injured by derailing cable car, 2/23/25
N. Yaworski	Mitchell explosion, Coast Range Tunnel, 7/17/30
Jose Zurano	Rock rolled from hillside at dam site, 4/25/20

Chief Engineer O'Shaughnessy in a 1924 speech said that 17 men working for the Utah Construction Company were killed in 3½ years of dam construction. The San Francisco *Chronicle* at time of project completion in 1934, states that 89 lives were "checked off" as the price of the project. M. M. O'Shaughnessy himself might be included on this list: he died of a heart attack on October 12, 1934, just a few days before the completion ceremony.

The author is indebted to Carlo M. De Ferrari, County Clerk and Auditor-Controller at Sonora, and Tuolumne County historian, for most of the information in this list.

Typical of interesting machinery used in Hetch Hetchy construction is this stiff-legged derrick placing pipe on the Moccasin penstock line in 1924. The derrick dragged itself up the long hillside, riding on skids atop the big pipes.

APPENDIX 2

HETCH HETCHY RAILROAD STEAM MOTIVE POWER—OWNED

No.	Type	Cylinders	Drivs.	BP	Weight On Driv.	Weight Total	Net TF	Builder	Notes
1	2-T Heisler	17 x 14″	40″	160	114,000 lbs.	114,000 lbs.	21,820	Heisler 1105/1906	Former Dempsey Lbr. #1. Sold 1923 to Pickering Lbr.
2	3-T Heisler	17 x 15	38	200	150,000	150,000	29,100	Heisler 1369/1918	Sold 1923, Pickering. At Traveltown, Los Angeles.
3	2-8-2	21 x 24	46	165	131,000	167,000	31,330	Baldwin 35780/1910	Former Youngstown & Ohio R. #1; sold 1927, Cal. & Ore. Coast #301.
4	2-8-2	20 x 28	48	180	144,000	190,000	33,200	A-Schen. 61535/1920	Sold 1927, Newaukum Vy. Ry. #1000; later SMV 1000. At Traveltown.
5	2-6-2	18 x 24	48	180	100,000	127,000	24,050	A-Cooke 62965/1921	Sold 1935, Weyerhaeuser #100.
6	3-T Shay	13½ x 15	36	200	198,000	198,000	35,100	Lima 3170/1921	Sold 1926, Pickering #6; retired 1958. At Yosemite museum.

Steam Locomotives Leased: Mt. Tamalpais & Muir Woods Shay #7; Sierra Ry. Shay #10 and #12; Northwestern Pacific Shay #251.

Steam Locomotives Used in Construction: Rolandi #7, 4-6-0 (former McCloud R.); Rolandi #9, Shay (former California Western); leased Sierra Ry. Shays #10 and #11.

GASOLINE-ENGINED MOTIVE POWER

—	Cadillac Railcar, no number. 7-passenger inspection car.			
6	Packard track truck, probably 1917 model.	3½-ton capacity.		Sold 1922.
7	Pierce-Arrow ″ ″ ″ ″ ″	2-ton, 16 passengers. Fitted as fire engine, 1922.		Gone 1929.
8	Pierce-Arrow ″ ″ ″ ″ ″	2-ton, freight.		Sold 1922.
9	Packard ″ ″ model GK.			
—	White ″ ″ reconverted to road vehicle in 1922. (Possibly No. 10)			
19	White track bus/ambulance.	12-passenger, ¾-ton.	Body by Meister.	Preserved Yosemite museum.
20	White track bus. 45 h.p.	23-passenger, 2-ton.	Ex-Desmond Park Co. #8.	Scrapped 1949.
21	″ ″ ″ Meister body.	27-passenger, 2-ton.	Rebuilt to track truck, 1925-26.	Scrapped 1949.
22	″ ″ ″ ″ ″	Express-mail, 2-ton.	Wrecked 1922.	Chassis used for #24.
23	″ ″ ″	24-passenger, 2-ton.	Ex-Nev-Cal-Oregon Ry.	Scrapped 1949.
24	Chassis from #22 rebuilt, HHRR.	32-passenger, 2-ton.	Body by Meister.	Demotored 1927; scrapped 1949.
1170	(Former 2nd #3) Plymouth loco.	8-ton (7½) Model DLC.	From Alturas 1934.	Sold 1949.
1171	(Former 2nd #2) Plymouth loco.	20-ton, Model HLA/2.	Plym. #3649/1931.	Became Parr Term. RR. #2.
1173	(Former caboose 16) Rotary snowplow built 1931, HHRR.		30 ft., 30-ton.	Sold 1949.
1176	(Former 2nd #1) Plymouth loco.	8-ton, Model DLC.	Plym. #2937/1928.	Sold 1949.
	Speeders used in "revenue" service included numbers 15-18, 28-30.			

In Addition: Steam loco. crane by Link-Belt; 40-ton steam shovel; 50-ton wrecker; wedge snowplow #101; five passenger-train cars; 41 freight cars (some dump cars lettered "City & Co. of S.F., H.H.").

BIBLIOGRAPHY

Books

Bean, Walton. *Boss Ruef's San Francisco.* Berkeley: Univ. of California Press, 1952.

Binger, Walter D. *What Engineers Do.* New York: W. W. Norton & Co., 1938.

Buckbee, Edward Bryan. *The Saga of Old Tuolumne.* New York: Press of the Pioneers, 1935.

California, State of. *Report of the Railroad Commission of California.* Sacramento: State Printing Office, various, 1918-1926.

———. *Review of Activities of the State Relief Administration of California, 1933-1935.* San Francisco: State Relief Administration, 1936.

———. *Water Law of California, 1921.* Sacramento: Division of Water Rights, 1921.

Carothers, Doris. *Chronology of the Federal Relief Administration, May 12, 1933, to December 31, 1935.* Washington: U.S.W.P.A., 1937.

Cooper, Erwin. *Aqueduct Empire.* Glendale: Arthur H. Clark Co., 1968.

Deane, Dorothy Newell. *Sierra Railway.* Berkeley: Howell-North Books, 1960.

Federal Power Commission. *Rules and Regulations Governing the Administration of the Federal Water Power Act.* Washington (First Revised Issue), 1921.

Freeman, John R. *The Hetch Hetchy Water Supply for San Francisco* (Freeman Report). San Francisco: Board of Supervisors, 1912.

Gudde, Erwin G. *California Place Names.* Second ed. Berkeley: Univ. of California Press, 1960.

Hammond, John Hays. *Autobiography of John Hays Hammond.* New York: Farrar & Rinehart, 1935.

Hanna, Phil Townsend. *Dictionary of California Land Names.* Los Angeles: Automobile Club of Southern California, 1946.

Harroun, Philip E. *Report to the Water Committee of the East Bay Cities on Water Supply for the Cities of Oakland, Berkeley, Alameda, and Richmond.* San Francisco, 1920.

Holmes, Roberta E. *The Southern Mines of California; Early Days of Sonora Mining Region.* San Francisco: The Grabhorn Press, 1930.

Jackson, Joseph Henry. *Anybody's Gold.* New York: D. Appleton-Century, 1941.

———. *Bad Company.* New York: Harcourt, Brace, 1949.

Jones, Holway R. *John Muir and the Sierra Club; The Fight for Yosemite.* San Francisco: Sierra Club, 1965.

Lewis, Oscar. *High Sierra Country.* New York: Duell, Sloan and Pierce, 1955.

Newport, Thomas L., ed. *The Lure of the Mark Twain-Bret Harte Counties.* Jamestown: *Mother Lode Magnet,* 1931.

Noble, John Wesley. *Its Name was M.U.D.* Oakland: East Bay Municipal Utility District, 1970.

O'Shaughnessy, Michael M. *Hetch Hetchy; Its Origin and History.* San Francisco: M. M. O'Shaughnessy, 1934.

Outland, Charles. *Man-Made Disaster; the Story of St. Francis Dam.* Glendale: Arthur H. Clark Co., 1963.

Sanchez, Nellie Van de Grift. *Spanish and Indian Place Names of California; Their Meaning and Their Romance.* San Francisco: A. M. Robertson, 1930.

Schlichtmann, Margaret, and Paden, Irene. *The Big Oak Flat Road . . .* San Francisco: Emil Schlichtmann, 1955.

Schussler, Herman. *The Water Supply of San Francisco, California, Before, During and After the Earthquake of April 18, 1906, and the Subsequent Conflagration.* San Francisco: Spring Valley Water Co., 1906.

Sierra Railway Co. of California. *In Old Tuolumne . . .* N.p., n.d. (c. 1899).

Spring Valley Water Co. *The Future Water Supply of San Francisco: A Report to the Honorable the Secretary of the Interior and the Advisory Board of Engineers of the United States Army.* San Francisco: Spring Valley Water Co., 1912.

Stoddard, Thomas R. *Annals of Tuolumne County.* Edited by Carlo M. De Ferrari. Sonora: Tuolumne County Historical Society, 1963.

Taylor, David W. *Life of James Rolph, Jr.* San Francisco: Committee for Publication of the Life of James Rolph, Jr., 1934.

Taylor, Ray W. *Hetch Hetchy; the Story of San Francisco's Struggle to Provide a Water Supply for Her Future Needs.* San Francisco: R. J. Orozco, 1926.

Tuolumne County, California: Being a Frank, Fair and Accurate Exposition, Pictorially and Otherwise, of the Resources and Possibilities of This Magnificent Section of California. Sonora: *Union Democrat,* 1909.

Winther, Oscar O. *Via Western Express and Stagecoach.* Stanford: Stanford Univ. Press, 1947.

Articles, Pamphlets, and Allied Materials

"Argument Against the Selection of Hetch Hetchy as a Source of Water Supply for San Francisco, Based Upon Report of George S. Nickerson, Consulting Engineer, Respecting the Availability of the South Eel River and Putah Creek Water-Sheds As a Source of Supply." Submitted by Henry M. McDonald to members of the United States Senate, November 25, 1913.

Brower, David R. "Footnote to Hetch Hetchy." *Sierra Club Bulletin,* vol. 39, no. 6 (June 1954).

Cogswell, Dr. L. J. "Early-Day Incidents of Every-Day Occurrence in Tuolumne County." *Grizzly Bear* (NSGW), vol. 8, no. 2 (Dec. 1916).

Cutler, Robert K. "Hetch Hetchy – Once is too Often." *Sierra Club Bulletin,* vol. 39, no. 6 (June 1954).

Dunn, Russell L. "Review of City Engineer Marsden Manson's Plans and Estimates of Hetch Hetchy Water Supply for the City of San Francisco." San Francisco, 1908.

Eckart, Nelson A. "Benefits Accruing From the Hetch Hetchy Project, San Francisco Water Supply." *Journal of American Water Works Assn.*, vol. 28, no. 9 (Sept. 1936).

Eddy, Elford. "The Building of Hetch Hetchy" (series of articles). San Francisco *Call,* San Francisco *Call and Post,* Dec. 1921-Feb. 1922.

Fries, Virginia. "The Hetch Hetchy Railroad." Unpublished manuscript, 1970.

Galloway, John D. "A Comparison of the San Pablo Reservoir of the East Bay Water Co. with the Calaveras Reservoir of the Spring Valley Water Co. as Sources of Water for the Cities of the East Side of San Francisco Bay." San Francisco: Galloway & Markwart, 1917.

———. "San Francisco Water Supply – Study of the available water on the Tuolumne River to determine the amount of stored water necessary to maintain a uniform flow of 400,000,000 gallons per day." J. D. Galloway, C.E., May 1919.

Hender, Arthur C. "Sierra Railroad: Farewell to Steam." *Western Railroader,* vol. 18, no. 6 (Apr. 1955).

"Hetch Hetchy Railroad." *Western Railroader,* vol. 24, no. 10 (October 1961).

"The Hetch Hetchy Story." Script, radio program. San Francisco: KSAN, Apr. 25, 1947.

"Hetch Hetchy Water Supply Edition." Sonora *Banner,* Mar. 19, 1920.

Lloyd, Harry E. "The Cherry River Project of the City of San Francisco." Mimeo., n.p., Aug. 17, 1949.

McAfee, Lloyd T. "Hetch Hetchy Project Nears Completion." Berkeley: Univ. of California *California Engineer,* vol. 12, no. 8 (May 1934).

Manson, Marsden. "Efforts to Obtain a Water Supply for San Francisco from the Tuolumne River." San Francisco, 1907.

Moore, Oral L. "Thirty Years of Hetch Hetchy Water." Speech, League of California Cities, Sept. 30, 1964.

Muir, John. "Everyone Help to Save the Famous Hetch-Hetchy [*sic*] Valley and Stop the Commercial Destruction Which Threatens Our National Parks." N.p., Nov. 1909.

Muir, John. "The Hetch Hetchy Valley." *Sierra Club Bulletin,* vol. 1, no. 4 (Jan. 1908).

"Old Railcar Makes Its Last Trip" (Hetch Hetchy Railroad). Modesto *Bee,* Apr. 25, 1965.

O'Shaughnessy, M. M. "Construction Progress of Hetch Hetchy Water Supply of San Francisco, California." *ASCE Bulletin,* vol. 48, no. 2 (Feb. 1922).

O'Shaughnessy, M. M. "The Hetch Hetchy Water Supply of the City of San Francisco." *Journal of American Water Works Assn.*, vol. 9, no. 5 (Sept. 1922).

O'Shaughnessy, M. M. "San Francisco's Hetch Hetchy Water Project." San Francisco *The Municipal Record,* Oct. 1934. Written two days before his death.

Phelan, James D. "The Hetch Hetchy and San Francisco." *Out West,* vol. 1, no. 32 (Feb. 1911).

Richardson, Elmo R. "The Struggle for the Valley; California's Hetch Hetchy Controversy, 1905-1913." *California Historical Society Quarterly,* vol. 38, no. 2 (Sept. 1929).

San Francisco, City of. "The $10,000,000 Hetch Hetchy Water Bond Issue." Chamber of Commerce, Sept. 1924.

San Mateo Historical Assn. Spring Valley Water issue of *La Peninsula,* vol. 14, no. 3 (Oct. 1967).

Stocker, Leslie W. "Some Engineering Features of the Enlargement of the O'Shaughnessy Dam." *Journal of American Water Works Assn.*, vol. 27, no. 8 (Aug. 1935).

Sullivan, Eugene J. "The Sierra Blue Lakes Water & Power Company's Offer to San Francisco." San Francisco, 1911.

Vilas, Martin S. "Water and Power for San Francisco from Hetch Hetchy Valley in Yosemite National Park." Reprint from San Francisco *Chronicle,* 1915.

Wurm, Ted. "Hetch Hetchy Railroad." *Chispa* (quarterly of Tuolumne County Historical Society), Apr.-June, 1970.

Wurm, Ted. "The 'Short Line' to Yosemite." *Chispa,* Apr.-June, 1970.

Publications and Records of the City and County of San Francisco

Board of Supervisors:

Official Document. "Report on Water Supply for the Sierra Nevada Mountains by the Special Committee on Water Supply of the Board of Supervisors of the City and County of San Francisco. Received and Adopted, October 8, 1906."

"Reports on the WATER SUPPLY of San Francisco, California, 1900 to 1908, Inclusive." Published by authority of Board of Supervisors, 1908.

"Statement to Voters Concerning a Proposition to Acquire a MUNICIPAL WATER SUPPLY As Submitted at the Special Election, Thursday, November 12, 1908."

Bureau of Engineering of the Department of Public Works:

O'Shaughnessy, M. M. "The Hetch Hetchy Project of the City and County of San Francisco: Its Progress, Prospect and Possibilities." And various similar titles, annually, 1920 to 1931.

"Recent Construction Activities." Pamphlet, Aug. 13, 1921.

"Report of the Bureau of Engineering of the Department of Public Works, City and County of San Francisco." Annual Report Books, 1909 to 1931.

San Francisco Office Diary, Dec. 1912 through 1917.

"San Francisco Water Rights on the Tuolumne River" (Book). Hetch Hetchy Water Supply: Robert M. Searls, Special Counsel, Dec. 1925.

Simkins, William A. "Report on Area Penetrated by the Aqueduct of the Hetch Hetchy Water Supply . . . – Foothill Division." 1934.

Stocker, Leslie W. "Report on the (Proposed) Operation of the Hetch Hetchy Railroad." Typescript, City Engineer's Office, Jan. 1917.

Various Maps, Drawings, Photograph Records, and Files at Department of Engineering, Hetch Hetchy Library, Utilities Engineering Bureau, Moccasin Powerhouse.

Public Utilities Commission:

Bartell, M. J. "Chronology of Hetch Hetchy Project." Jan. 1939.

Bartell, M. J. "Population Forecast for Northern California and the San Francisco Bay Region." Apr. 1936.

"Plan Nine for Municipal Distribution of Electric Power." Aug. 1941.

"San Francisco Water and Power" Booklet. Various years, 1935 to 1949.

"Raker Act (H.R. 7207)." Reprinted by Public Utilities Commission, Sept. 1941.

"Report of the San Francisco Public Utilities Commission." Annual Report Books, 1932 to 1970.

Wehe, Roy A. "Hetch Hetchy Water and Power Division, Separation Study and Power Division Cost and Rate Analysis for City and County of San Francisco." 1955.

Wurm, Ted, ed. "San Francisco Water and Power." 1967.

Various Files at Public Utilities Commission photo laboratory.

Newspapers

(San Francisco Unless Otherwise Indicated)

Bay Guardian, Bulletin, Call, Call & Post, Call-Bulletin, Chronicle, Daily Transcript (Oakland), *Enquirer* (Oakland), *Examiner, Mother Lode Magnet* (Jamestown), *News, Post Enquirer* (Oakland), *Tribune* (Oakland), *Tuolumne Independent* (Sonora), *Tuolumne Prospector* (Groveland, later Jamestown), *Union Democrat* (Sonora).

Periodicals

ASCE Bulletin
California Minute Man
The City (San Francisco)
Commonwealth Club Transactions (San Francisco)
Electric Light & Power
Electrical West
Engineering News (and *Eng. Record, News-Rec.*)
Journal of American Water Works Assn.
Journal of Electricity (various titles)
Municipal News
P.G.&E. Progress
Pony Express
Power Plant Engineering
Public Utilities
Public Works
Railway Age (Gazette)
Railway and Locomotive Engineering
Railway Mechanical Engineer
Railway Review
San Francisco *Municipal Record*
San Francisco Water
Scientific American
Sierra Club Bulletin
Sunset
Timberman
Western Water News

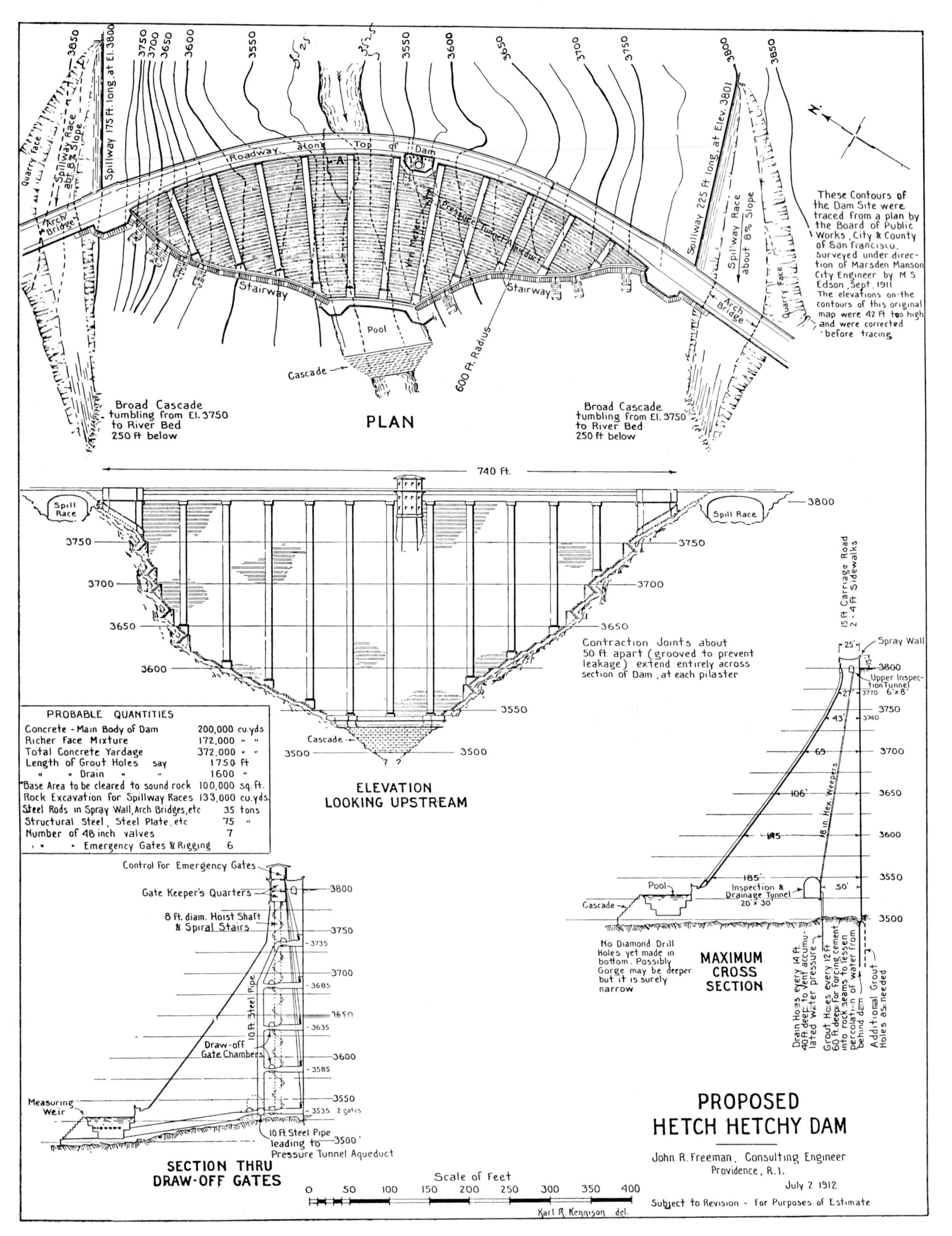

One of several proposed designs for the dam at Hetch Hetchy, this appeared in the "Freeman Report" of 1912. Fortunately, it was considerably changed by Chief Engineer O'Shaughnessy before construction.

INDEX

*Denotes picture reference

– A –

Ackerson Meadow, 33
Adit 5-6 Camp, 108*, 112
Adit 8-9 Camp, 110*, 113
Aggregates plant, 91*, 100, 223*, 227, 252*, 255*, 257*
Alameda County, 14, 16, 229, 230, 238*
Alameda Creek, 16, 209, 216, 218*, 223*, 227, 229
Alameda East Portal, 223*
Alameda West Portal, 221
American River, 19, 21, 24
American Society of Civil Engineers, 43, 142*
Aquatic Park, 12*, 13
Armstrong, Burl, 84*
Armstrong "Shoveloder," 109*
Arndt, Fred, 132*, 144
Arroyo Del Mocho, 215*. *See also* Mocho Shaft
Arroyo Del Valle. *See* Valle Shaft
Aston, Taggart, 26
A.T.&S.F. (Santa Fe) R.R., 63, 169, 172, 204*

– B –

Baird, Charlie, 18, 19, 36*, 45, 136, 210
Baird Hotel, 38, 129*
Baker, Jim, 57, 67, 260
Baker, Mose, 59*
Ballinger, Richard A., 22
Bank of America, 211, 222
Bank of Italy, 211*. *See also* Bank of America
"Bartel-Manson Report," 28
Bartell, Max J., 19*, 22, 28, 41, 68, 209, 218*, 219, 222, 224*, 226, 230, 273, 277*, 282
Bartlett, George, 23*, 25
Baseball. *See* Groveland baseball
Battery locomotives, 106*, 198, 213, 214*, 221
Bay Cities Water Co., 17, 21, 22, 24*
Bay Crossing Division, 229-235*
Bay Crossing Division (map & profile), 228
Bay Crossing Pipeline, 228
Bay-Pulgas Pumping Station, 235
Beck, Ernie, 6, 137, 157, 166, 176, 245
Bendel, Bill, 68
Bensley, John, 13
Berkeley Camp, 169, 171, 249
Berkeley, City of, 17
Bernhard, Waldo, 53, 57, 59, 61
Best, Jack, 48, 67, 113, 209, 222
Biddle, Col. John, 22, 25
Big Creek, 33, 60
Big Creek Camp, 110*, 111*
Big Creek Shaft, 111*, 114*, 148
Big Creek trestle, 54*
Big Oak Flat, 29, 32*, 33*, 45, 53, 247*
Big Oak Flat Road, 25, 29*, 31*, 32*, 33, 34*, 41, 44*, 45, 51*, 186, 191*, 245, 249, 273
Big Oak Lodge, 249
Black Point, 12*, 13
Blagg, Sam, 36*
Blanchard, Will, 133*
Blue Lakes, 14, 20*, 26
Boitano, Leonard, 57
Boothe, Frank, 48, 105
Bootlegging, 135-137
Boyle, T. F., 84*
Brown Adit, 194*, 195,* 196*, 199
Brown Adit Tramway, 149, 194*, 199
Brown, George, 34*, 37*
Buck Meadows (Hamilton Station), 25, 52*, 60, 112, 132, 137, 162, 163*, 249
Bus competition, 167
Byington, Lewis, 261, 266*

– C –

Cable incline. *See* Incline railways
Cableway, overhead, 87*, 101*, 149, 197*, 199, 201, 241, 257, 260
Cahill, Edw. J., 226*, 266
Calaveras Dam, 16, 229
Calaveras Reservoir, 216, 230
Calaveras Valley, 16
Calderwood, Shorty, 112
California & Oregon Coast R.R., 275*
California Industrial Safety Dept., 221, 227
California Peach Growers (and Railroad), 144, 145*, 146*, 147*, 176, 177
California Peach Growers R.R. (map), 168
California, University of, 22, 28
California Weekly, 22

California-Western R.R. & Navigation Co., 58
Callaghan, Joe, 224*
Canyon Power Tunnel, 259, 277, 279*, 280*
Canyon Ranch, 72*, 73, 78*, 82*
Caplinger, Earl "Cappy," 157, 176
Carbona, 213
Carl Inn, 134*
Carle, Jackson, 217*
Carne, Ray, 70, 132*, 135, 157, 166, 169
Case, Jack, 64*, 132*, 144
Cassaretto, Floyd, 131*
Cassaretto, Mrs. Fred, 131
Cavagnero, Joe, 57, 260
Cavagnero Siding, 186
Cement, 86, 94, 100
Central Valley Project, 266
Chabot, Anthony, 13
Chaffee, Horace B., 86, 139, 185*; Mrs. Chaffee, 66*
Charlotte Hotel, 137
Chase, Hal, 130
Cheminant, Bob, 132*, 157
Cheminant, Edith, 132*, 159
Cheminant, Lester B., 63, 70, 121, 132*, 159, 172, 175
Cherry Dam, 277, 278*, 280*
Cherry Lake, 280*
Cherry Penstock, 279*
Cherry River (Creek), 21, 73, 79, 269, 270, 277*
Cherry Valley, 21, 26
Cherry Valley Reservoir, 277
Chinese Camp, 32*
Chinese Station, 30*, 33, 40*, 42*
Clear Lake, 14
Cleary, Al J., 63, 84*, 121
Clerk, Charles, 172
Cliff House, 30
Coast Range, 209
Coast Range Tunnel, 209-227, 230; completion, 226*; criticism, 227; map, profile, 208*; safety, 227
Colburn, Roy, 133*
Colfax Springs, 54
Collard, Earl "Hap," 130
Colman, Jesse, 238
Concrete plant, 94*, 95*, 97*, 99*, 100
Connolly & De Luca, 199
Connolly, Tom, 57, 132
Conservation, conservation groups, 19, 22, 191
Construction camps (general), 105. *See also* individual camp names
Construction Co. of North America, 118, 121
Consumnes River, 21, 24
Conty, Mary, 263
Cooper, J. D., 48
Corral Hollow, 213, 223*
Corral Hollow Pipeline, 219*, 220*, 221, 222
Cosby, Maj. Spencer, 22
Cosmopolite Mine, 29
Coulterville, 263
Coulterville-Yosemite Road, 19
Cowan, Mike, 132*
Coyle, Edmund, 133*, 175*
Crane Flat, 25
Criddle (Superintendent), 113
Crockers (Crocker Station), 31*, 33, 45*
"Crookedest Railroad in the World," 58
Crowley, Art, 266, 267*, 278*
Crystal Springs Dams, 13, 15*, 16*, 230*, 237*
Crystal Springs Reservoir, 229, 230*, 235, 238
Cummings, Frank, 187

– D –

Damsite Camp, 56*, 61, 102*, 176, 257
Dart, J. P., 16
De Ferrari, Carlo M., 285
De Ferrari, Charles, 133*
De Ferrari, Frank, 133*, 140*
De Martini, 115
Degnan, Doctor John, 117, 118*, 119, 130, 142, 216
Depression years, 23, 222, 241, 251
Derrick, stiff-legged, 283*
Dinky locomotives, 86, 87*, 91*, 148, 179, 181*, 182*, 186
Divide, 51*, 57
Dolge, William, 191
Don Pedro Dam, 197, 198, 277, 282
Don Pedro Reservoir, 50, 149, 197, 198, 277, 281*, 282
Donaldson, George C., 45
Drill sharpener, 112, 196*
Dumbarton Bridge, 232*, 233*
Dumbarton Strait, 235
Duran, Tom, 166, 172, 175
Durgan, Earl, 133*, 166

– E –

Eagle-Shawmut Mine, 29, 32, 37*
Early Intake, 60, 73, (road) 76*, 77*, 78, 79, 81*, 105, 241, 277. *See also* Intake Station
Early Intake Dam, 120*, 124, 125*, 241
Early Intake Incline, 80*, 81*, 105, 106, 145, 149*, 241, 242*, 249
Early Intake Powerhouse, 77*, 79, 80*, 124
Easler, Thornton, 243
East Bay Cities (E.B.M.U.D.), 85, 209, 218*, 221
Eckart, Nelson A., 63, 79, 152*, 218*, 224*, 226*, 230
Ecological considerations, 73, 82, 88, 261
Eddy, Edwin, 226*
Edison battery locomotives, 108
Eel River, 17
Eggert, Billy, 40*
Eleanor Creek, 21
Eleanor, Lake. *See* Lake Eleanor
Electric-battery locomotives. *See* Battery locomotives
Electric line crew, 77*
Electric power, 17, 73, 77*, 79, 80*, 124, 179, 191, 268*, 270. *See also* Power transmission line
Electric power disposition, 266
Electric trucks, 223*, 227
Electrical Development League, 169, 172
Ellis, Owen, 105
Elwell, Francis and Lewis, 30
Elwell-Parker electric truck, 223*, 227
Engineering News-Record, 100, 221, 238
Erskine, Ed, 144
Espey, T. W., 226*
Europa Hotel, 134*, 166

– F –

Feather River, 17
Femons, Hank, 246, 266
Femons, Harvey, 54*
Fent, Curtis "Red," 41, 67, 246, 247, 261, 266
Ferretti, Sal, 136
Filmer, George, 226*
Fisher, Walter L., 25
Fitzgerald, W. J., 84*
Fleming, Tom, 172
Flick, C. Bruce, 121
Flood control, 277
Flory hoist, 88*, 89*
Floyd, Marie, 132*. *See also* Neel, Marie
Foothill Division, 148, 149
Foothill Division Tunnel, 193-204; digging competition, 201
Ford, Tom, 117
Fort Mason, 12*
Fort Point, 13
Fowler, Superintendent, 114
Francisco Reservoir, 13
Fraser, D. G., 84*
Freeman, John R., 22, 25, 41, 47, 229
Freeman Plan, 43
"Freeman Report," 25, 26*
Freight teams, 32*, 33, 34*
Fry, C. H., 227
Fry, John, 135

– G –

Gallagher, Andy, 112
Galloway, John D., 17, 22, 66*
Gambling, 135-137
Gardiner, Curley, 127, 130
Garfield, James R., 21
Garfield Permit, 21, 22
Garrotte, 29. *See also* Groveland
Gartland, J. B., 84*
Gasoline rail vehicles (general), 151-157*, 213, 216
General Electric battery locomotives, 108, 117*
Giannini, Amadeo P., 211*, 222
Gillespie (State Inspector), 66*
Gilliam, Del, 177, 245, 249
Glaciers, 17, 18*, 85, 86
Godwin, Kenneth, 261
Gold mining, 29, 33, 263
Golden Gate, 13
Golden Rock Ditch, 33, 39, 83, 115, 124
Goldsworthy, Sy, Jr., 36*
Golub, Joe, 126*
Golub, Pete, 278*
Gould, Dr. Elisha, 83, 119
Graham, Jimmy, 136, 216
Grahame, ––, 132*
Gravel Spur Siding, 148*
Graves, Roy D., 265
Gray, Alvin, 195*
Gray, Roscoe, 36*
Greek Camp, 112
Grizzly Gulch, 33, 53
Groveland, 19, 33*, 34*, 38*, 41*, 43*, 44*, 45, 47, 57, 59, 60, 70*, 83*, 126*, 127, 128*, 129*, 130-138, 162, 260, 273. *See also* Hetch Hetchy Project Headquarters
Groveland (map), 168
Groveland baseball, 126*, 127, 130, 131*, 132
Groveland District, 39
Groveland fires, 127, 134*, 166
Groveland headquarters, 193, 196
Groveland hospital, 78*, 83*, 210
Groveland Hotel, 38*, 44*, 134*, 136, 137
Groveland Social Club, 132
Groveland water, 115, 127
Grunsky, Carl E., 17, 21, 41
Guthrie, A., & Co., 199

– H –

Haight, Catherine Cobden, 135
Haley, John "Mike," 117
Hall, Charlie, 36*
Hall, William Ham, 21, 47
Hamilton Station, 60. *See also* Buck Meadows
Hammond, John Hays, 21
Harder, Bob, 126*
Hardin Flat, 33, 73
Hardy, Bob, 132*
Harper, Charlie, 36*
Harte, Bret, 115
Harten, Ivan, 150*
Harwood, Julian, 54*, 68, 144, 261
Havenner, Franck, 238
Hawkins, Horace, 133*
Hayden, J. Emmett, 218*, 226*
Hazel Green, 19
Healy, Clyde, 28, 224*
Helbush, Bill, 227, 258, 260
Henniker, Henry, 34*
Hercules gasoline locomotives, 257
Hermann (engineer), 229
Hetch Hetchy, Hetch Hetchy Valley, 17, 18*, 19, 20, 21, 25, 26, 27*, 35*, 47, 62*, 72*, 73, 85, 89*, 90*, 91*, 92*, 169
Hetch Hetchy & Yosemite Valley Railroad, 41, 47
Hetch Hetchy Aqueduct, 47, 105, 193
Hetch Hetchy Aqueduct tunnel. *See* Mountain Division Tunnel; Foothill Division Tunnel; Coast Range Tunnel; Pulgas Tunnel
Hetch Hetchy Bay bridge, 233*, 235*
Hetch Hetchy Bond Purchase Syndicate, 222
Hetch Hetchy bonds, 43, 192*, 210*, 211, 222
Hetch Hetchy dam. *See* O'Shaughnessy Dam
Hetch Hetchy Department, 225, 227
Hetch Hetchy grant. *See* Raker Act
Hetch Hetchy Junction, 46*, 49*, 53, 71*, 149, 153*, 175, 193, 196, 202*, 204, 260*, 264*, 265*, 270
Hetch Hetchy Junction (map), 192
Hetch Hetchy Junction school, 201*
Hetch Hetchy Junction Shaft, 199, 201, 202*
Hetch Hetchy Lodge, 169
Hetch Hetchy opposition, 9, 17-28, 229, 283
Hetch Hetchy Power Division, 204, 240, 243, 254, 263
Hetch Hetchy Project, 17, 40, 47, 73, 85, 209; commendation, 238, 239; completion, 236*, 237*, 238; financing, 9, 21, 41, 43, 84, 85, 119, 121, 193, 220, 222, 251; future, 85; headquarters, 83*, 129*, 132*; map sketch, 11; priorities, 43, 45, 209; quality, 209, 283; success, 239; tunnel vs. pipeline, 209

Hetch Hetchy Railroad, abandoned, 224*, 271; accidents, 170*, 172, 173*, 174*, 175, 245*; construction, 47, 53, 55*, 56*, 61, 148; deterioration, 243, 251, 270; dismantling, 224*; electrification, 63; employees, 159; end of common-carrier service, 176; equipment disposition, 196, 239, 271; equipment roster, 286; operation, 63, 67, 139-177, 241-249; operation by Sierra R.R., 251-260; rehabilitation, 254; right of way, 52*, 61, 63, 67, 243, 251, 269, 270, 282; station list, 165; timetable, 155, 156, 164*
HHRR ambulance. *See* Track bus No. 19
HHRR bridge, 160*
HHRR cement trains, 159
HHRR enginehouse-shop, 68*, 70*, 126*, 133*, 138*, 142, 175*, 176*
HHRR gasoline vehicles, 65*, 66*, 68, 246, 286. *See also* HHRR track buses; HHRR track trucks; HHRR inspection car; HHRR speeders; HHRR snow fighting and equipment
HHRR Gravel Spur, 52*
HHRR inspection car (Cadillac), 66*, 67*, 68, 71*, 152*
HHRR inspection trains, 172
HHRR locomotives, 286; Heislers, 67, 69*; leased, 63, 138*, 172; Mikados, 67, 196; No. 1, 58*, 67, 69*, 275*; first No. 2, 67, 69*, 275; second No. 2, 246 (*see also* No. 1171); No. 3, 71*, 139, 140*, 143*, 160*, 161*, 170*, 275*; No. 4, 139, 141*, 143*, 163*, 172, 173*, 174*, 275; No. 5, 139, 141*, 162, 163*, 170*, 171*, 174*, 177*, 196, 204, 276*; No. 6, 133*, 139, 141*, 143*, 167*, 175, 196, 273, 276*; No. 1170, 240*, 261, 269*, No. 1171, 240*, 260, 261, 271*, 272*; No. 1176, 240*; preserved, 275
HHRR Moccasin Branch, 183, 204
HHRR passenger trains, 162, 166, 167, 169-172, 176, 249, 260. *See also* HHRR track buses
HHRR passes, 164*
HHRR railfan trips, 265*, 269*, 270*, 271*, 272*
HHRR rolling stock, caboose, 67, 245*, 246; disposal, 196, 239, 271; freight, 52*, 63, 148*, 250*; passenger train, 142*, 169, 171*, 253*, 261, 271, 272*
HHRR shop. *See* HHRR enginehouse-shop.
HHRR snow fighting & equipment, 68, 158*, 160*, 166*, 167, 242, 245-49, 260, 265*, 269*, 270; rotary snowplow, 245*, 246, 247*, 248*, 249, 254, 260, 263, 265*, 273, 278*; snowplow No. 101, 166*, 167, 246
HHRR special trains, 169-172, 263
HHRR speeders, 61*, 70, 152*, 155, 176, 242*, 246, 269*, 270*, 271
HHRR tie tamper, 54*, 67
HHRR track buses, 151-157, 169, 196; No. 19, 65*, 151, 153*, 155, 157, 242*, 243, 246, 271, 273*; No. 20, 150*, 151, 153*, 204, 245; No. 21, 151, 153*, 155, 242*, 245, 246, 256*, 272*; No. 22, 151, 154*, 175; No. 23, 151, 156*, 157, 245; No. 24, 151*, 155, 156*, 242*, 245, 246, 256*, 263, 269*
HHRR track trucks, 64*, 66, 68, 70, 154; Packard, 64*, 152*; Pierce-Arrow, 64*, 66*, 245
Hetch Hetchy Reservoir, 88*, 259*, 282*
Hetch Hetchy spirit, 6, 10, 133, 201, 283
Hetch Hetchy survey, 40*
Hetch Hetchy workmen, 201
Hetch Hetchy Valley. *See* Hetch Hetchy
Hetch Hetchy Valley resort, 19
Hetch Hetchy Valley camp, 90*
Hewitt, Dixwell, 84*
Hile, Louis, Mr. & Mrs., 78*, 126
Hitchcock, E. A., 19
Hoar mucking machine, 108, 122
Hodeau Flat, 73
Hog Ranch, 18*, 19, 47, 60, 82. *See also* Mather
Holm Powerhouse, 277, 281*
Holmes, Bert, 133*
Hope, Earl, child, 118*
Hope, Phil, 137
Hope, Tom, 133*, 142, 266
Horse transportation, 18*, 22, 23*, 30*, 31*, 32*, 34*, 36*, 37*, 82*, 88*
Hospital, Groveland, 117, 119, 210, 245; Livermore, 210
Hotchkiss (engineer), 166, 167
Hughes Station and Ranch, 186
Hull's Meadows, 47
Humphrey, W. F., 121

– I –

Ickes, Harold L., 238, 266
Incline railways, 71, 73, 79, 80*, 81*, 86, 89*, 105, 106, 121*, 140, 145, 148, 149*, 180*, 197
Indian Creek Shaft, 213, 217*, 225*
Intake station, Intake siding, 56*, 79, 81, 241, 269*
Irrigation districts, 19, 207, 273, 277. *See also* Modesto I.D.; Turlock I.D.; Waterford I.D.
Irvington, Irvington Portal, 209, 216, 224*, 230, 234, 235
Isaacs, Mary. *See* Meyer, Mary

– J –

Jacksonville, 29, 38*, 53, 130, 263, 271, 281*
Jameson, Leonard "Slim," 193
Jamestown, 29, 254
Jerome, Black Jack, 122
Johns, Andy, 79, 187, 188*
Johnson, Matt, 107
Jones, Calvin, 166
Jones, Drenzy, 41
Jones Station, 55*, 60
Journal of Electricity & Western Industry, 191
Jubb, Milo, 133*, 140*, 142

– K –

Kennedy, Harley, 161*, 166
Kings River, 14
Kirchen, Herman, 140
Kirkwood Powerhouse, 277, 281*
Knowland, Joseph, 43
Kolano Rock, 27*
Kyte, Al, 132

– L –

Labor problems, 121, 122
Laetus, Julius, 117
Laguna Honda Reservoir, 13
Lake Bigler, 14
Lake Eleanor, 17, 21, 26, 47, 72*, 73, 75*, 82, 145, 277, 283
Lake Eleanor Dam, 73, 75*, 76*, 79
Lake Eleanor road, 258*, 259*, 260
Lake Merced, 13
Lake Tahoe Water Co., 14
Lane, Franklin K., 26
Lamm, Tex, 117
Lawlor, Hank, 89*
Lawrence, Vince, 126*, 132, 133*
Lawron, William, 30
Laveroni, Bernice, 130, 133
Laveroni, George, 137, 195*
Laveroni, Jake, 36*, 133*
Lichenstein, Joy, 84*
Lidgerwood cableway. *See* Cableway, overhead
Livermore, Livermore Valley, 210, 213, 229, 266
Lobos Creek, 13
Logging, 82, 149
Lombard Reservoir, 12, 13
Long, Byrd, 133*
Los Angeles Aqueduct, 229
Lower Cherry Aqueduct, 74*, 76*, 77, 79, 124, 148, 241, 243*, 244*
Lumsden Bridge, 47
Lumsden, Tom, 133*, 140*

– M –

McAfee, Lloyd, 63, 89*, 121, 142, 172, 176, 217*, 225, 226, 238, 256*, 261
McAtee, Lou, 60, 113, 115, 121, 126*, 127, 130, 195*, 204
McCarthy, D. S., 77*
McCarthy, Eugene, 36*
McCarthy, John A., 121
McCloud River survey, 19*
McDougald, John, 84*
MacDougall, Don, 216
McIntosh, R. P., 66*, 86
McKenna, Tom, 166, 172
McLeod, W. S., 117
McMann, C. C., 77*
McLeran, Ralph, 84*, 124
Magee, Irene, 124
Magee, Walt, 127, 132, 133*, 139, 175*
Mallett, Everett, 78*
Manson, Marsden, 16, 21*, 22, 25, 26, 41, 47
Marin County, 13
Marion steam shovel, 48*
Marsh's Flat, 196
Mather, Camp Mather, 79*, 82, 146*, 147*, 161*, 169, 176, 177, 247, 248*, 249, 256*, 257, 260. *See also* Hog Ranch
Mather (map), 168
Mather, Stephen T., 82
Medical facilities. *See* Hospital
Meikle, Roy, 273, 277*, 283
Meister, A., & Sons, 65, 151, 154
Mercado, Louis, 41
Mercer, Max, Mr. & Mrs., 78*
Merrill, Millard, 136
Meyer, Gene, 112, 119, 246
Meyer, Mary, 117, 119
Miguel Meadow, 257, 261
Miller, Frank, 117
Minard, Bert, 132, 133*, 175*
Minton, H. A. (architect), 186
Mirk, Doug, 70, 78, 268
Mirk, Muriel, 78*
Mission Dolores, 13
Mission San Jose, 209, 216
Mitchell Shaft, 212*, 213*, 214*, 216, 227
Mitchell, Shorty, 77*
Mitchell, Walt, 245, 246
Miwok Indians, 17, 33
Moccasin, 148, 193, 204, 241, 246, 249, 251, 263, 264*, 266; dams, 189*; incline railway, 185*, 186, 263; map, 178; penstock, 182, 183, 184*, 185*, 186, 187*, 188*, 285*; portal, 194*, 210; powerhouse, 45, 180, 184*, 186*, 187*, 189, 190*, 191*, 243, 250, 268*, 277, 280*; reservoir, 188*, 189*
Moccasin Creek, 29, 45, 53, 178, 186, 188*, 189*, 282
Mocho Shaft, 213, 214*, 215*, 217*, 227
Modesto Irrigation District, 19, 269, 273, 277
Mokelumne River, 14, 21, 26, 28, 221, 283
Moore, Oral L., 277
Morena Dam, 41
Morris, Shorty, 166
Morse, Shorty, 162
Mother Lode, 115
Mother Lode League (baseball), 127
Mt. Carmel Church, 133
Mt. Jefferson Mine, 29, 33*
Mount Lyell, 17
Mt. Tamalpais & Muir Woods R.R., 58*, 59*
Mountain Division camps, 105
Mountain Division Tunnel, 101*, 104*, 105, 107*, 108, 112, 113*, 118*, 119, 121, 122*, 124*, 277
Mountain Lake Water Co., 13
Moxom, Marshall, 259
Moyer, Ethel, 119, 157
Moyles, J. W., 84*
Muheim, Emile, 40*
Muir, John, 20*, 22, 26, 28
Mulholland (engineer), 229
Munn Tank, 162
Myers-Whalley mucking machine, 108, 109*, 122

– N –

Nagel, Herman, 166, 177
Narrow-gauge rail lines, 73, 80*, 86, 145, 179, 195*, 216, 218*, 223*, 224*, 241, 243*, 257. *See also* Valley Railroad; Incline railways; Brown Adit Tramway
Needy, Walter, 126*, 132
Neel, Jim & Marie, 132, 144, 187
Nevada-California-Oregon Ry., 151
Newark, 262*
Newark-San Lorenzo pipeline, 218*, 221
Newark Slough, 234*, 235
Newaukum Valley R.R., 275*
Newell, William H., 53, 251
Newport, Clara, 144
Newspaper attacks, 85
Nielson, Bob, 36*
Nolan, Shorty, 136
Northwestern Pacific R.R. locomotive No. 251, 138*, 172
NRA program, 251

– O –
Oakdale Portal, 196, 204
Oak Flat Toll Road, 45. *See also* Big Oak Flat Road
Oakland, 43, 209, 221
Oakland *Daily Transcript*, 14
Oakland Recreation Camp, 39*, 55, 124*, 156*, 169, 171*, 249
Oakland *Tribune*, 43, 176
O'Brien, Willis, 127, 130, 132, 144, 221
Ocean Shore R.R., 151
Ohmer, Henry, 224*
O'Keefe, Mickey, 161*, 166
Old Mission Portland Cement, 118*, 121
Oliver-Harriman Mine, 29
O'Shaughnessy Dam, completion (initial), 88, 102*, 103*, 171, 172; completion (after raising), 258, 259*, 260, 261; concrete placing, 87; construction, 85, 92*-99*; diversion tunnel, 88*, 98*; foundation, 93*, 95*, 96*, 99*; raising, 250-264
O'Shaughnessy, Michael M., "Chief," 6, 9, 25, 26, 28, 84*, 88, 89*, 102*, 124, 127, 157, 205*, 209, 218*, 224*, 226*, 238, 273, 285*; character, 193, 238; cooperation with irrigation districts, 207; death of, 235, 237; demotion, 225, 226; devotion, 193, 229, 238; employed by San Francisco, 41; labor attitude, 122; memorial, 260; opposition to, 210; speeches, 43, 57*, 63, 193, 221
Ost, Paul, 79, 224, 226, 243, 254
O'Toole (City Attorney), 273

– P –
Pacific Gas & Electric Co., 243, 262, 266, 270
P.G.&E. *Progress*, 229
Pardee, George, 218*
Parr Terminal R.R., 271
Peach Growers. *See* California Peach Growers
Pechart, Walter "Peach," 136-137
"Pecker Point," 137
Pedro Adit, 149, 196, 206*
Pedro Siding, 162, 172
Pelton water wheels, 186, 190*, 268*
Peninsula (area), 13, 14
Peninsula cities' water supply, 230, 283
Peninsula population study, 230
Perkins, Senator, 43
Perry, George, 121
Peterson, Pete, 112, 215*, 217*, 226
Peterson, "Sealskin Pete," 190*
Phelan, Eddie, 57
Phelan, Mayor James, 17, 21, 22
Phelan, Les, 157, 176, 246, 249, 278*
Phelan Ranch, 130, 131*
Phillips, Joe, 66*
Pickering Lumber Co., 275*, 276
Pickett, Mr. & Mrs., 161*
Pilarcitos Creek, Reservoir, 12*, 13, 15*
Pillette, Ed, 133*
Pipelines. *See* Bay Division Pipeline; San Joaquin Pipeline; Corral Hollow Pipeline; Newark-San Lorenzo Pipeline
Plymouth locomotives, 195*, 199, 216, 240*, 241, 245, 246, 254, 257, 271
"Poison Oakers," 29
Politicians, 209, 238
Polk, Willis, 229
Pool, Earl, 57
Poopenaut Pass, 60, 158*, 163*
Poopenaut Valley, 33
Power transmission line, 243, 262, 266, 267*, 268*, 269*, 278
Pracy, George W., 226*
Prewitt, George, 133*, 140*, 142
Priest Dam, 116, 148, 179, 180*, 181*, 182*
Priest Hill (railroad grade), 47, 53*, 57, 162, 176
Priest Hill Road (Priest Grade), 53*, 57, 162, 176
Priest Portal, 116*, 117*, 123*, 179, 180*
Priest Portal camp, plant, 116, 121*
Priest Reservoir, 62, 116, 148, 179, 182*
Priest's, Priest Hotel, 38*, 53, 134*
Prinz, Oscar, 68
Prohibition, 172
Prostitution, 135-137
Public ownership of utilities, 9, 15, 16, 17, 22, 230, 239, 266, 270; opposition, attacks on, 14, 17, 19, 22, 191, 229
Pulgas Tunnel, 229, 231*, 235
Pulgas Water Temple, 236*, 238
Purdy Company, 271
Putah Creek, 14

– Q –
Quast, Joe, Mr. & Mrs., 132*

– R –
Raggio, Dave, 57
Railway Age, 63
Raker Act, 28, 176, 221, 266, 270, 282, 283; conditions imposed by, 19
Raker, John E., 28
Rancheria Mountain, 27*
Rankin, Carl R., 48, 60, 216, 217*
Rasmussen, Jim, 133*, 138*
Rathbun, Harry, 126*, 132
Rattlesnake Gulch, Creek, 33, 53*, 57, 62*, 162*, 180, 181*, 182*
Rawles, Ernie, 116
Red Hill, 176, 242*, 246, 253*
Red Mountain Bar, 47, 148, 149, 196, 197*, 198*, 199*, 201
Redwood City, 229, 231, 235
Reed, Bill "Busher," 130, 132, 283
Reese, Frank, 190*
Reese, Roy, 133*
Republican mines, 29
Riordon, Tim, 84*
Riverbank, 205, 269
Road vehicles (motor), 78*, 79, 104*
Rock River, 196, 201, 202*
Rodgers, Merle, 124, 245
Rolandi construction trains, 51*, 57, 59*
Rolandi, Frederick S. (contractor), 46*, 48*, 53, 57*, 60, 61
Rolandi, F. S., Jr., 61
Rolandi locomotives, No. 3, 57; No. 7, 46*, 51*, 57; No. 9, 50*, 57, 58*; No. 10, 50*, 57, 59*; No. 11, 57, 61*
Rolph, Mayor James, Jr., 28, 41*, 84*, 88, 102*, 103, 172, 211, 218*, 225

Roosevelt, F. D., 251, 266
Rosasco, 47, 196
Rose, Al, 271, 273
Rose, Dr. Homer, 119
Rossi, Mayor Angelo, 226*, 238, 261
Rubicon River, 14
Ruef, Abe, 22
Ryan, John "Buddy," 115, 121, 216, 226, 257, 260
Ryan, Mickey, 263

– S –

Sacramento River, 17
Salvage, 239
San Andreas Fault, 14, 16
San Andreas Valley, 13
San Antonio Reservoir, 229, 238*
San Francisco Bay, 229, 230, 232*, 233*, 234*, 235*
San Francisco Bay Area, 16; map, 228
San Francisco Bay Pipeline, 16. *See also* Bay Crossing Pipeline
San Francisco *Bulletin,* 21
San Francisco *Call-Bulletin,* 238
San Francisco *Chronicle,* 229, 254, 285
San Francisco, City & County of, 6, 13; Board of Public Works, 17; City Engineer, 28; Department of Engineering, 221; graft scandal, 21, 22; Grand Jury, 185*; "new" charter, 17, 222, 224; Public Utilities Commission, 221; sued by U.S. Attorney General, 266; Supervisors, 14, 18*, 19, 21, 22, 23*, 71*, 209; Water Department, 13, 229, 230, 236*, 239; Water Works, 12*, 13
San Francisco earthquake & fire, 9, 19, 28, 40*
San Francisco *Examiner,* 23, 226
San Francisco Midwinter Fair, 41
San Francisco Municipal Ry., 66, 68, 70, 151
San Francisco Municipal Record, 220*, 238
San Francisco water, quality, 17, 239; rates, 283; search, 13-22, 26*
San Joaquin Pipeline, 203-207
San Joaquin River, 14, 16, 29, 203*, 204, 205*, 207
San Joaquin Valley, 29, 211*
San Lorenzo, 218*
San Mateo County, 12*, 13, 15*
San Mateo Creek, 13
Santa Clara County, 16
Santa Maria Valley R.R., 275*
Sawmills, 72*, 73, 78*, 79*, 82
Schader, A. D., 271
Schmitz, Mayor Eugene, 21, 22, 24
Schott brothers, 132, 133*, 175*
Schussler, Herman, 14
Scott, Bob, 126*, 144
Scowden, T. R., 14
Scarles, R. M., 66*, 84*
Seco Substation, 210
Seco Summit, 221
Second Garrotte and Camp, 31*, 115*
Segale, Angelo, 117
Segale Store, 263
Selby, Mayor, 14
SERA. *See* State Emergency Relief Act
Seward, Carl, 278*
Shaw, Ben, 73, 82*
Shay locomotives, 139, 143. *See also* HHRR locomotive No. 6; Sierra Railway
Sheafe's Hotel, 38*
Sierra Blue Lakes Water & Power Co., 20*, 26
Sierra Club, 9, 22, 26, 28, 283
Sierra Nevada, 16
Sierra Railway (Railroad), 30*, 33, 46*, 47, 48, 49, 63, 157, 162, 167, 175; dispatching, 254; finances, 254, 261; locomotive No. 10, 50*, 57, 59*; locomotive No. 11, 57, 61*; locomotive No. 12, 139, 183; locomotive No. 20, 253*, 260*; locomotive No. 34, 253*, 261*; locomotive No. 36, 260; operation of HHRR, 251-261
Sinclair, Neil, 166
Six Bit Gulch, 48*, 50*, 58*, 144*, 162, 173*, 174*, 254
Six Bit Gulch trestle, 172, 198
Sladden, Dick, 117
Slade Gulch, 37*
Sloan, Al, 166
Smith, J. E., 166
Smith, J. Waldo, 238
Smith Station (Smith's), 113, 161*
Smoot, H. R., 84*
Snider, Dud, 78*, 246, 249, 254, 260
Snider, Margaret Golub, 196, 201*
Sno-Cat, 278*
Snook, Stanley, 271
Snow King plow, 246
Sonora, 29
Sonora baseball team, 130
South Fork, 30, 45, 169, 249, 251, 260, 277; camp, 55, 67, 104*, 105, 106*, 107*, 108, 112, 123*, 124; trestle, 56*; station, 55*, 64*, 120*
South Fork Lumber Co., 145
Southern California Mountain Water Co., 41
Southern Pacific R.R., 48, 63, 167
Speedway engine, 246, 247
Spring Valley Water Co., 13, 14, 16, 19, 22, 23, 25, 209, 229, 230
Stagecoaches, 18, 30*, 32*, 33, 45*, 186
Stanislaus National Forest, 29
State Emergency Relief Act (SERA), 241, 250, 251, 254
Steam shovel, 48*, 53*
Steengrafe, Milt, 130
Stenander, Harry, 142, 153*
Stevens Bar, 263, 281*
Stocker, Leslie W., 41, 86, 226*
Sunol, 213
Sunol Dam, 16
Sunol Water Temple, 229
Sunset District wells, 221
Sunset Reservoir, 236*

– T –

Taft, President, 22
Taylor, Lt. Col. Harry, 22
Taylor, Ray, 84*
Tesla Portal, 204, 207, 211*, 213
Thomas Shaft, 211*, 213, 222, 223*, 227
Tinkler, Charlie, 121
Tomasso, Leonardo, 56*
Townley, Ranger, 25
Townsend, Andy, 77*, 187, 188*
Townsend, Don, 77*
Townsend, Orland, 187, 188*
Tracy, 204
Tramway, aerial. *See* Cableway, overhead
Tramway, cable. *See* Incline railways

Transbay Construction Co., 251
Trencher, Buckeye, 203*, 205*, 207
Tunneling work & methods, 105, 113*, 114*; accidents, 117, 216; concreting, 118*, 120*, 122*, 123, 200*, 201, 215*, 224*, 227, 231*; railroads & machinery, 104*, 106*, 107, 108, 109*, 213; rescue squad, 213*, 216; workmen, 108, 112, 114, 116, 201, 215*. *See also* Mountain Division Tunnel; Foothill Division Tunnel; Coast Range Tunnel; Pulgas Tunnel
Tuolumne City, 47, 130
Tuolumne County, 18*, 21, 29-39; employment of residents, 127, 251; roads, 33, 45, 114
Tuolumne Electric Co., 39
Tuolumne Prospector, 39
Tuolumne River, 9, 10, 16, 17, 19, 22, 25*, 73, 85, 98*, 106*, 198*, 259*, 263, 273, 281*, 283; Canyon, 50, 51*, 53, 60*, 197*; Middle Fork, 36*, 264*; South Fork, 32*, 33, 54*, 105
Tuolumne River Power Co., 33
Tuolumne Water & Supply Co., 21
Turlock Irrigation District, 19, 269, 273, 277, 283
Turner, James, 205*, 225, 226, 277, 283
Turner, James, Dam, 229, 238*
Tuttle, Howard, 48
Tyler, J. G., 84*
Tzars, Enrique, 117

– U –

U.S. Army engineers, 22, 25
U.S. Congress, 19, 25, 28
U.S. Dept. of Interior, 266
U.S. Geological Survey, 16, 21
U.S. Mail, on HHRR, 167, 176
U.S. Public Works Administration, 261
U.S. Supreme Court, 266
Universal Concrete Gun Co., 122
University of San Francisco, 210
Upper Alameda Creek Tunnel, 230
Utah Construction Co., 47, 84, 85, 86, 87*, 89*, 285; rail equipment, 86, 87*, 90*, 91*, 92*, 95*-98*

– V –

Valle Shaft, 213, 219*, 227
Valley Railroad, 86, 87*, 90*, 91*, 95*-98*, 100*, 145

– W –

Walsack, Hilda, 132*
Wanderer, William "Red," 193, 207, 249
Ward, Bert, 271
Ward's Ferry Bridge, 282
Warnerville Substation, 277
Warren, George, 87
Water depletion & claims, 124, 201, 204, 227
Water rights, 249
Water Resources Center Archives, 28
Water shortages, in S.F., 9, 14, 23, 26, 218, 220*, 221, 222, 229
Water sources map, 26
Water Works & Sewerage, 239
Waterford Irrigation District, 19
Watson, Ed, 36*
Wattis, W. H., 84*, 85, 89*
Webb concrete gun, 122, 123, 227
Webb, Eddie, 263, 264*, 265, 266
Webb, W. F., 122
Wehner, A. J., 121, 217*
Western Pacific R.R., 213
Westside Lumber Co., 47, 277
Weyerhaeuser Timber Co., 276*
Wheeler, Benjamin Ide, 22
Wheeler, Louie, 226*
Whitcomb locomotive, 216, 224*
White, Charlie, 36*
White, Frank, 77*
White, Jack, 36*, 124, 127, 130, 132, 133
White Motor Co., 273. *See also* White trucks, road; HHRR track buses; HHRR gasoline vehicles
White, Ray, 67, 172
White, Susie, 132, 133
White trucks, road, 78*, 220*, 246, 269*, 270; rail, 151
Wilbur, Ray Lyman, 266
Wiley, Bill, 115
Wilson, Lincoln, 144
Wilson, President Woodrow, 26, 28
Wood, R. J., 86
Woods Creek, 29
Works Progress Administration (WPA), 251

– Y –

Yosemite National Park, 16, 17, 19, 20*, 25, 29, 130, 132, 221, 261
Yosemite National Park Museum, 273, 276
Yosemite Park Co. (railroad operation), 169
Yosemite Power Co., 33, 73, 83
Yosemite Short Line R.R., 33, 39*, 47, 53
Yosemite tourists, 167, 169; on HHRR, 151
Youdall Construction Co., 205*, 207
Yuba River, 17

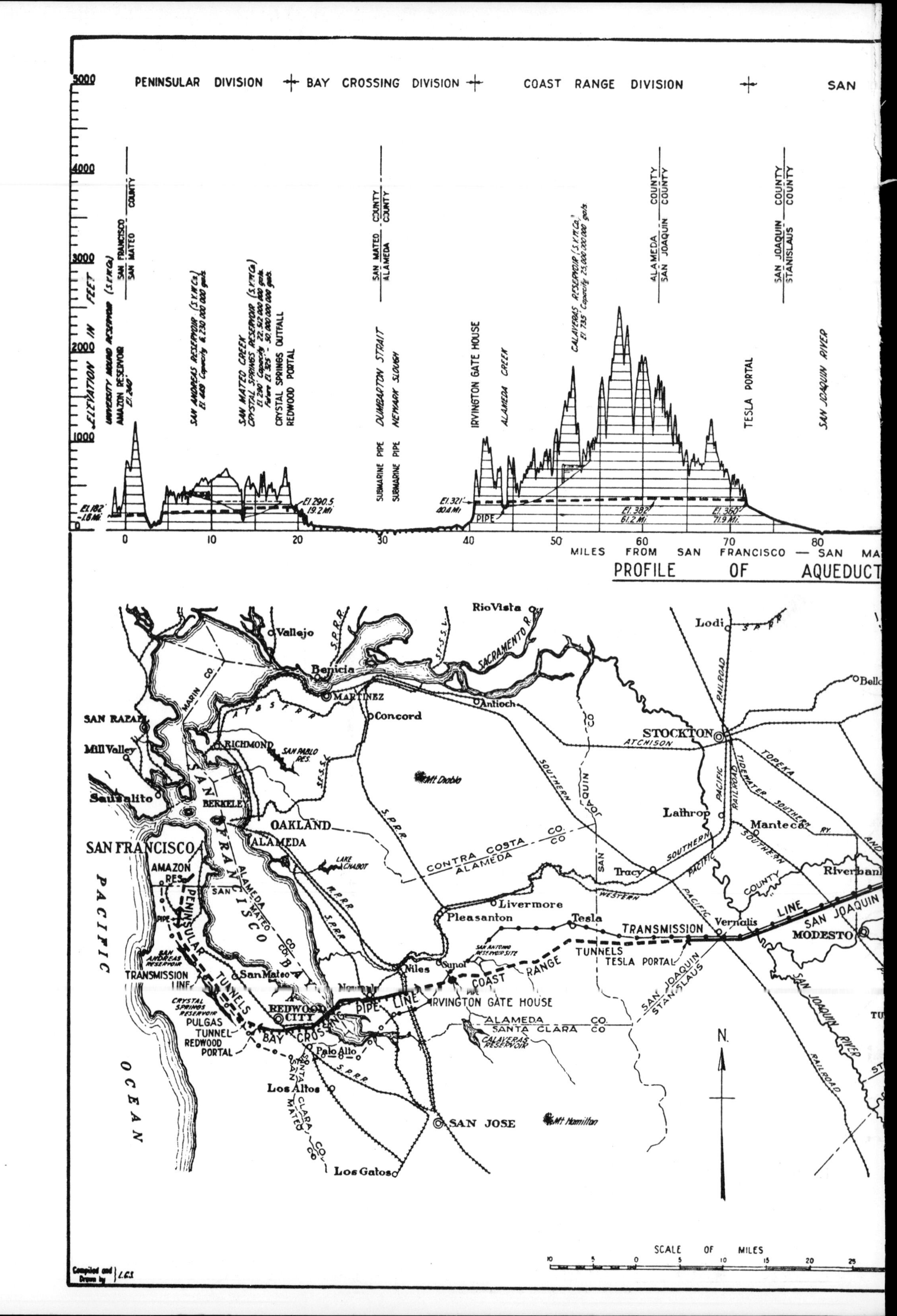

PENINSULAR DIVISION
BAY CROSSING DIVISION
COAST RANGE DIVISION
SAN
ELEVATION IN FEET
MILES FROM SAN FRANCISCO — SAN
PROFILE OF AQUEDUCT
TRANSMISSION LINE
COAST RANGE TUNNELS
TESLA PORTAL
IRVINGTON GATE HOUSE
PIPE LINE
BAY CROSSING
PENINSULAR TUNNELS
PACIFIC OCEAN
SAN FRANCISCO BAY
SCALE OF MILES
Compiled and Drawn by L.E.S.